INTECH

open science | open minds

Contents

Preface

Humanity is facing several critical global challenges at the beginning of the 21st century. One of which includes the quest for alternative energy resources that mitigate the dependence on fossil fuels. Whereas fossil fuels are available in situ at all times, the utilisation of renewal energies has to cope with large temporal fluctuations ranging from seconds to seasons. The passing shadow of a cloud over solar panels causes the fastest variability of power output followed by the gustiness of the wind, the rise and fall of the tides and the seasonal and annual variations of the availability of biological resources for energy generation. Thus, the kinds of questions being asked of the research community have changed over the last decades, reflecting the increasing awareness of the finite nature and the instability of fossil fuel supply.

Capturing wind energy has been widely employed for centuries – i.e. the traditional windmills of the Netherlands being a significant landscape element for centuries. To date, the emerging market for wind power energy is experiencing remarkable global growth rates which affect not only the problem of how to technically link these into existing power systems, but also effect deeply rural landscapes and local livelihoods. In many instances, initial positive local acceptance altered to the contrary, leading to sometimes strong opposition against the instalment of wind turbines and wind farms in rural landscapes. Hence, solving this problem requires additional input of economists and social-political scientists. The emerging interdisciplinary research increased the understanding and helped to develop adequate solutions to many of the problems revolving around wind power energy. However, the disciplinary integration and interdisciplinary understanding must be much further advanced.

This book is a timely compilation of the different aspects of wind energy power systems. It combines several scientific disciplines to cover the multi-dimensional aspects of this yet young emerging research field. It brings together findings from natural and social science and especially from the extensive field of numerical modelling.

Harvesting wind power requires the erection of towers with rotating wings in the landscape or at sea. Such artificial buildings with moving parts modify drastically the natural views of the panorama. This raises the question of what are the initial necessary societal preconditions and attitudes to erect a wind turbine. Furthermore, new grid codes are needed that addresses the requirements to allow the integration of the variable power generated by renewable energy systems into existing power networks. Several contributions discuss issues revolving around the variable power of a single wind turbine, which poses high demands on its control, and means of buffer storage. These technical aspects and problems are enhanced for clusters of turbines in a wind park and the complexity of safe power transmission of large and variable power from wind farms over long distances. The book then moves beyond the classical wind farm aspects and explores potential spin-off products of offshore wind farms. A case in point is the potential of creation of new marine habitats. Various aspects of making a secondary use of the rigid offshore wind turbine basement constructions as anchor device for aquaculture in the open ocean is discussed in the final chapter.

The actual research questions of the societal challenges raised in this book should be not only framed and articulated by scientists but more and more with policy makers and relevant stakeholders, particularly those concerned with adaptation strategies and sustainable development.

However, one of the major struggles remains how to further define, develop and implement integrative research that studies, explains and projects the various interactions within human-environment and renewable energy systems. Although the book does not provide cast-in-stone solutions to the critical challenges, it outlines the science needed to address these challenges in the near future. Thus, a better understanding of manifold dimensions of wind energy systems is the core aim of this book.

Dr. Gesche Krause
Social Science and Coastal Management
Leibniz Center for Tropical Marine Ecology (ZMT), Bremen,
Germany

Part 1

Introduction

Local Attitudes towards Wind Power: The Effect of Prior Experience

Jacob Ladenburg[1] and Gesche Krause[2]
[1]AKF, Danish Institute of Governmental Research
[2]ZMT, Leibniz Center for Tropical Marine Ecology
[1]Denmark
[2]Germany

1. Introduction

Globally the market for wind power energy is experiencing some of the largest growth rates in history. New markets are emerging and existing markets are expanding. The rural landscape as we know today is thus changing as new, larger and more efficient turbines are erected.

At a general level the positive public acceptance of this change in the landscape appears to be associated with where people have been consulted prior to the instalment, thus acknowledging the potential local opposition towards specific projects. However, several imperative questions remain open. Does the social acceptance and wind power development go hand in hand? Or will the large increase in the wind power capacity have negative repercussions on the attitudes? And if so, is the change in attitude dependent on specific types of characteristics of the wind turbine development, which people gain experience with?

In order to reduce potential negative feedback mechanisms from wind power development on attitudes a look into the "crystal ball" would be helpful. If we *ex ante* can foresee some of the most obvious caveats associated with wind turbine development, we might be able to apply anticipatory planning that may mitigate the negative effects from wind turbine development on the acceptance of wind power. Fortunately, existing attitude surveys contain information, which can be employed to assess how attitude and wind power development will be related in the future. Many of the existing attitude wind power studies have included variables, which account for different types of "experience" that the respondents in the surveys had with wind turbines. These variables often entail information on whether or not the respondents have had a "physical/visual" experience with wind turbines, such as a view to turbines from the residential property, distance to turbines, number of turbines in the local area etc. In Denmark, these prior experience variables represent people who are living in a landscape with more wind turbines than the general population. By examining the attitudes of these specific groups of respondents, we are able to shed light on how attitudes may alter in future landscapes with higher levels of wind turbine densities.

The present chapter therefore provides a review of these studies and discuses the results in relation to what can be "glimpsed" in the crystal ball for the future social acceptance of wind power generation.

2. A model for prior experience

Attitude formation towards wind power is far from being straightforward and clear predictions are not easy. The central theme in most of the wind power literature focusing on public attitudes is how turbines under different settings and circumstances can generate opposition (see Gross 2007; Graham et al. 2009; Jobert et al. 2009; Jones & Eiser 2009; Ladenburg 2009 and Haggett 2011 for some of the more recent papers on this subject).
As mentioned, the aim is to shed light on how prior experience with wind turbines might have an influence on attitude, and most importantly how we relate this to attitudes towards the wind farm landscapes to come in a near future. We therefore need a model that takes prior experience into account. Quantitative analyses of attitudes in previous research on wind power attitude formation and prior experience typically define differences in individual attitudes in a linear form, in which individual i's latent attitude q_i^* is a function of the individual demographics, X_i, and a set of variables, θ_i, representing one or several dimensions of prior experience with wind turbines, see below

$$q_i^* = X_i\beta + \theta_i\varphi \ . \quad (1)$$

In the assessment of the influence of prior experience, θ_i, is thus the cornerstone variable. As we will see in the review, θ_i represents different types of prior experience definitions, which depend on the information available in the studies. This typically includes information on whether the respondent lives near a wind turbine or has seen a turbine, i.e. during a general visual encounter. However, some of the studies also have more detailed information, such as distance to the wind turbines, number of turbines seen on a daily basis or systematic differences in the experience with wind turbines. All in all, these variables cover different types of experience with wind turbines, which a larger share of the total population of a country will experience in the coming years' wind power landscapes.
Consequently, the information from existing studies provide insights on how attitudes can develop (all things else being equal) if more people in the future are exposed to these types of experiences.
That said, prior experience is only analysed here in the frame of visual/physical encounters with actual wind turbines. Naturally, physical encounters are just one source of prior experience. Prior experience or perhaps more correctly prior information can be obtained from various numbers of sources. Compared to prior experience, which refers to a personal experience, prior information can be obtained through indirect experience, such as relatives´ or friends´ experience with wind turbines and their expressions thereof. Prior information can also be obtained by reading positive or negative articles in the newspaper, watching the news in television and through other types of media (Kuehn 2005; AMR interactive 2010). This is important to keep in mind when interpreting the results from this review and when we relate these to the wind turbine landscapes in the near future.

3. Review

As stressed in the prior experience model, the prior experience relates to a "physical" encounter with wind turbines entails different types of information. In the review below, these differences are addressed and related to the type of wind power development location, i.e. general attitude towards wind power and attitudes towards specific locations of development. Hereby land-based and offshore installations are distinguished.

To increase the accessibility of the review and the subsequent results, the studies obtained from the literature are categorised according to the type of location, i.e. land-based or offshore systems. Within each category, the studies are presented in chronological order by the first author. Besides the name of the study, the table lists, which prior experience variables were included, whether the variables were significant and in that case the direction of the effect (positive or negative). In this relation, "<0" should be read as the prior experience variable having a negative effect and ">0" as having a positive effect on the stated attitude. If the effect of the variables is marked as ^{NS}, this denotes that the effect is not significant at a 90% level of confidence. A * denotes significance at least at a 90% level of confidence.

3.1 General attitudes towards wind power

Several studies in the literature do not specifically address attitudes towards on-land or offshore wind farms, but elicit attitudes towards wind power in a broader context. Some of which are presented in table 1.

One of the first studies that addressed this issue was the paper by Krohn & Damborg (1999). Based on a Danish study, they reported from a survey carried out in a local area with many turbines. It was found that the distance to the nearest wind turbine and attitude are invariant. Accordingly, the distance to the nearest on-land turbine does not seem to have an influence on the attitude. Indeed they found that respondents who could see between 20-29 turbines from their home and who were living within 500 m from the nearest wind turbine tend to be more positive towards wind power in general. Unfortunately, they did not indicate whether these results were statistically significant.

Study	Focus of the paper	Prior experience variables	Effect of the variables
Krohn & Damborg (1999)	Attitude towards wind power	Living less than 500 m from existing turbines	$\beta_{Distance}{}^{NS}$
		Number of turbines visible from the residence of the respondents	$\beta_{No.\ turbines\ visible}{}^{NS}$
Ek (2005)	Attitude towards wind energy	Living near turbine(s)	$\beta_{Near\ turbines}{}^{NS}$
Meyerhoff et al. (2010)	Perception of the environmental quality	Number of encounters with wind turbines in the past four weeks	$\chi^2_{Wind\ turbine\ encounters}{}^{NS}$

Table 1. General wind power studies that focus on attitude and prior experience (compilation based on Ladenburg & Möller (2010)).

In a Swedish study (Ek 2005) it was tested if respondents who live near wind turbines have a different attitude towards wind power compared to respondents who do not live near turbines. The analysis could not establish such a connection ($\beta_{Near\ turbines}{}^{NS}$).

In Germany, Meyerhoff et al. (2010) analysed if there are any significant relationships between the number of wind turbine encounters during the last four weeks and the individual satisfaction of the regional environmental quality. Controlling for daily encounters, repeatedly encounter, encounter 2-3 times per week, only one encounter and no encounters at all, they find no significant differences in satisfaction. Environmental quality in the region thus appears to be independent of the number of encounters with wind turbines.

3.2 Attitudes towards land-based turbines

Table 2 comprises a list of studies that analyse the potential relations between attitudes towards on-land turbines and prior experience with wind turbines.

Focusing on attitudes, and the local intention to oppose turbines, Johansson & Laike (2007) tested in a Swedish study if residential prior experience variables related to the distance to the local turbines and to the view on these turbines. None of the variables are found to be significant in influencing individual perception and possible opposition.

In a Danish study by Ladenburg (2008) the attitude towards more on-land turbines based on a survey from 2003-2004 was analysed. The study included two experience variables, e.g. whether the respondent could see on-land or/and offshore wind farms from the permanent/summer residence. The results suggest that only in the case that the respondent can see both on-land and offshore wind turbines ($\beta_{\textit{View On-Land and Offshore}}$) prior information seems to influence the attitude towards more on-land turbines. In this particular case, prior information has a negative influence. Accordingly, respondents who have both an on-land and offshore wind farm in their view have a more negative attitude towards the prospect of a further increase of land-based turbines compared to respondent who either do not have a wind turbine in the view shed or have an on-land or offshore wind farm in the view shed from the permanent/summer residence.

In a following study, Ladenburg & Dahlgaard (2011) asked respondents about the attitude towards the existing on-land wind turbines. The relationship between attitude and prior experience were analysed by using information on whether the respondent could see on-land or/and offshore wind farms from the permanent/summer residence and the perceived number of wind turbines that each respondent sees on a daily basis. In addition, interactions between having a view shed to a wind turbine and the number of wind turbines seen on a daily basis were also tested. The test of the effect of prior experience showed that having a wind turbine in the view shed did not influence the attitude. Respondents who could see an on-land or/and an offshore wind turbine from their permanent/summer residence were equally positive/negative towards existing on-land wind turbines as the respondents who did not have a wind turbine in the view shed. Interestingly, the number of turbines seen daily had a significant effect on the attitude. More specifically, Ladenburg & Dahlgaard (2011) showed that respondents who see more than 5 turbines/day ($\beta_{>5 \text{ turbines per day}}$) have a more negative attitude compared to respondents, who see fewer turbines (0-5 turbines/day). Among the respondents who see 6-10, 11-20 or more than 20 turbines each day, attitudes are not significantly different between the respondents.

Based on the same data set as Ladenburg & Dahlgaard (2011), Ladenburg et al. (2011) analysed in a complementary study whether the number of land-based wind turbines seen on a daily basis affects the attitude toward more on-land wind turbines. The analysis suggests that having more than 20 turbines in the local area has a significant negative influence on the attitude towards more on-land turbines ($\beta_{> 20 \textit{ turbines per day}} < 0$). Indeed, they found that the relation between attitude towards more on-land wind turbines and the number of turbines seen on a daily basis is dependent on whether the respondents have a view to on-land turbines or not from the residence. More specifically, the respondents who have an on-land wind turbine in the view seem to be highly sensitive towards the number of turbines seen daily ($\beta_{6\text{-}20 \text{ turbines per day} | \text{on-land turbine in the view shed}} < 0$ and $\beta_{> 20 \text{ turbines per day} | \text{on-land turbine in the view shed}} < 0$. Furthermore, the negative effects seem to be increasing with the number of turbines seen daily, $\beta_{6\text{-}20 \text{ turbines per day} | \text{on-land turbine in the view shed}} < \beta_{> 20 \text{ turbines per day} | \text{on-land turbine in the view shed}}$. If the respondent do not have a wind turbine in the view shed, those who saw between 0-5 wind turbines per day were equally positive/negative as the respondents who saw more than 20 turbines, $\beta_{6\text{-}20 \text{ turbines per day} | \text{no on-land turbine the view shed}}^{NS}$ and $\beta_{>20 \text{ turbines per day} | \text{on-land turbine in the view shed}}^{NS}$).

Study	Focus of the paper	Prior experience variables	Effect of the variables
Warren et al. (2005)	Attitude towards two existing wind farms	Distance from residence to wind farm	$\beta_{Distance}^{*}$ (significance, see text)
Johansson & Laike (2007)	Intention to oppose additional wind turbines	Living at different distances from existing wind turbines	$\beta_{Distance}^{NS}$
Ladenburg (2008)	Attitude towards more on-land turbines	View to on-land turbines from permanent residence or summerhouse	$\beta_{View\ on\text{-}land}^{NS}$ $\beta_{View\ on\text{-}land\ and\ offshore}^{*} < 0$
Ladenburg & Dahlgaard (2011)	Attitude towards existing on-land turbines	View to on-land turbines from permanent residence or summerhouse	$\beta_{View\ On\text{-}land}^{NS}$
		View to offshore turbines from permanent residence or summerhouse	$\beta_{View\ Offshore}^{NS}$
		Number of on-land turbines seen on a daily basis	$\beta_{See\ more\ than\ 5\ turbines/day}^{*} < 0$
Ladenburg et al. (2011)	Attitude towards more on-land wind turbines	View to on-land turbines from permanent residence or summerhouse	$\beta_{View\ On\text{-}land}^{NS}$
		View to offshore turbines from permanent residence or summerhouse	$\beta_{View\ Offshore}^{NS}$
		Number of on-land turbines seen on a daily basis	$\beta_{See\ more\ than\ 5\ turbines/day\ |\ on\text{-}land\ turbine\ in\ view\ shed}^{*} < 0$

Table 2. Attitude and prior experience towards land-based turbines (modified from Ladenburg & Möller (2010)).

Warren et al. (2005) conducted two surveys on attitude towards existing and planned on-land turbines in two local regions in Scotland and Ireland. Focusing on the Irish study, the attitudes towards two specific wind farms in both Cork and Kerry were cross tabulated with the distance (0-5 km, 5-10 km and 10-20 km) from the residence of the individual respondent to the wind farms. The attitude frequencies point towards that the closer the respondents live to the wind farms, the more positive they are. Using the frequencies from the Warren et al. (2005), Ladenburg & Möller (2010) tested and confirmed these findings. With regard to the attitude towards the first established wind farm in the respective area, respondents living between 0-5 km and 5-10 km from the wind farms (one in Kerry and one in Cork) have similar attitudes. However, when comparing the attitudes between respondents living 0-5 and 10-20 km from the two wind farms, respondents living between 10-20 km from the wind farms were found to be significantly more negative.

3.3 Attitudes towards offshore wind farms

In the following section, the attitude studies focusing on offshore wind farms are presented (Table 3).

Study	Focus of the paper	Prior experience variables	Effect of the variables
Bishop & Miller (2007)	Perception of visual impacts from offshore wind farms at 4, 8 and 12 km from the shore	Location of on-land turbines in the neighbourhood	$\beta_{Turbines\ neighbourhood\_4\ km}{}^{*}<0$
			$\beta_{Turbines\ neighbourhood\_8\ km}{}^{NS}$
			$\beta_{Turbines\ neighbourhood\_12\ km}{}^{NS}$
Ladenburg (2008)	Attitude towards more offshore turbines	View to offshore turbines from permanent residence or summerhouse	$\beta_{View\ on\text{-}land}{}^{NS}$ $\beta_{View\ offshore}{}^{NS}$
Ladenburg (2009)	Perception of visual impacts from offshore wind farms	View to on-land turbines from permanent residence or summerhouse	$\beta_{View\ on\text{-}land}{}^{NS}$
		View to offshore turbines from permanent residence or summerhouse	$\beta_{View\ offshore}{}^{NS}$
Ladenburg (2010)	Attitude towards existing offshore turbines	Systematic differences in prior experience between two samples of respondents	$\beta_{Systematic\ differences}{}'<0$
		View to on-land turbines from permanent residence or summerhouse	$\beta_{View\ on\text{-}land}{}^{*}>0$
		View to offshore turbines from permanent residence or summerhouse	$\beta_{View\ offshore}{}^{NS}$
		Number of on-land turbines in the neighbourhood	$\beta_{No.\ turbines\ neighbourhood}{}^{NS}$
Ladenburg & Möller (2010)	Attitude towards existing offshore	Same as Ladenburg (2010) Travel time to the nearest	$\beta_{Traveltime}{}^{*}<0$ $\beta_{Traveltime2}{}^{*}>0$

turbines	offshore wind farm	$\beta_{Traveltime\_30min}{}^{*}<0$
	Number of turbines in the nearest offshore wind farm	$\beta_{Number\ of\ turbines}{}^{*}>0$
	Distance/height relation of the turbines in the nearest offshore wind farm	$\beta_{Distance/height}{}^{NS}$

Table 3. Attitude and prior experience towards offshore turbines (based on Ladenburg & Möller (2010)).

Bishop & Miller (2007) tested prior information in a study by analysing the visual impact from an 18 turbine offshore wind farm which could be viewed from the coast at different sites from the shoreline at 4, 8 and 12 km respectively. Prior experience was analysed using the approach described in Ek (2005)]. It was investigated whether respondents living in an area with land-based wind farms perceived the visual impacts from offshore wind farms as being more severe when compared to respondents without any contact with wind farms or living in an area with proposed or approved wind farms. The results suggest some influence of prior information, though the prior information effect was ambiguous. Apparently, this effect was found only to be significantly different in the case of visual assessment of the wind farm located at 4 km offshore, but not so, if the farm was viewed further way at 8 and 12 km off the coast.

Ladenburg (2007) analysed the attitude towards future offshore wind farms in Denmark. The paper includes prior experience information related to variables controlling for whether the respondents have a view shed to on-land and/or offshore wind turbines. The results suggest that prior experience does not influence the attitude towards future offshore wind farms ($\beta_{View\ On\text{-}Land}{}^{NS}$, $\beta_{View\ Offshore}{}^{NS}$).

However, in a complementary study, Ladenburg (2009) modelled prior information as a function of view to on-land and offshore wind farms from permanent residence or summer homes. This analysis combined controls for prior information by sampling respondents with distinctively different levels of experience with visual impacts from offshore wind farms. That was done by a selective sampling approach in which only respondents living close to Nysted I and Horns Rev I offshore wind farms along the Danish North Sea were sampled. The distinctly different levels of visual experience are obtain, as the offshore wind farms at Nysted and Horns Rev are located at approximately 6-9.5 km and 14-20 km off the shore, respectively. Whilst the wind farm at Nysted is very visible, the wind farm at Horns Rev is difficult to see during fair weather conditions due to the location far off the coast.

Analysing the prior experience variables (view on land based or offshore wind turbines) separately for the respondents from the Nysted and Horns Rev samples, the variables are not significant ($\beta_{View\ On\text{-}Land}{}^{NS}$ and $\beta_{View\ Offshore}{}^{NS}$). Thus, people who can see an on-land or offshore wind farm from their permanent or summer residence do not have a significantly different perception of the visual impacts compared to respondents who do not have a wind turbine in their view shed. However, when Ladenburg (2009) compared the perceptions of the visual impacts between the two sample locations, a strong prior experience effect seems to be present. More specifically, the results point towards that experience with relatively large visual impacts from offshore wind farms (Nysted I sample) has a rather negative influence on the perception of visual intrusion from offshore wind farms on the landscape, in contrast to people who have experience with offshore wind farms (Horns Rev I) with fewer/weaker visual impacts.

In Ladenburg (2010) the analysis of prior information was extended by including variables controlling for the perceived number of daily encounters with on-land wind turbines, i.e. number of turbines in the neighbourhood, where the respondents live. It was found that having a view to on-land turbines had a significantly positive influence on attitude ($\beta_{View\ On\text{-}Land}>0$). The respondents, who had an on-land wind turbine in the view, thus were more positively inclined towards offshore wind farms, compared to the respondents who did not. The number of turbines and view to offshore wind farms were not found to have any significant impact on the attitude.

In a final study, Ladenburg & Möller (2010) use the travel time from the residence to the nearest offshore wind farm as an indirect proxy for a prior experience in terms of a physical/visual encounter with the nearest offshore wind farm. Analysing the effect of prior experience on attitude towards existing offshore wind farms in Denmark, an ordered logit analysis suggests that the travel time has a significant influence on the attitude towards offshore wind farms. Generally, , the farther away the respondents live from one of the six offshore wind farms in this survey, the more negative are the respondents towards existing offshore wind farms ($\beta_{Traveltime}{}^{*}<0$), though at a decreasing rate ($\beta_{Traveltime2}{}^{*}>0$). However, the results denote that people living within 30 minutes of travelling to the nearest offshore wind farm are significantly more negative towards the offshore wind farms ($\beta_{Traveltime\_30min}{}^{*}<0$), suggesting some kind of negative proximity effect.

Controlling for the number of turbines and the distance/height relation (the smaller the distance/height relation is, the larger visual impacts and vice versa), it was also found that if the nearest wind farm contained many wind turbines, the respondents were more positive towards offshore wind farms ($\beta_{Number\ of\ turbines}{}^{*}>0$). However, the distance/height relation did not appear to have an influence on the attitude ($\beta_{Distance/height}{}^{NS}$).

4. Prior experience and implications for the future development of wind farms

In the previous sections, the potential influences of prior experience with wind turbines on the attitude towards different aspects of wind power development were presented. In this section, the results from this review are elaborated and discussed in relation to which information the studies indicate for the future development of wind farms. The discussion will focus specifically on the type of experience, such as number of turbines seen daily, having wind turbines in a view shed, etc.

4.1 Number of turbines

One of the fundamental wind power planning aspects is, how many wind turbines an area can contain without having too negative impacts on the local acceptance of wind power. Focusing on Ladenburg & Dahlgaard (2011) and Ladenburg et al. (2011), there seems to be a relationship between attitude towards land-based wind turbines in general and the total number of wind turbines in the local area. Apparently, higher numbers of turbines reduces acceptability of both existing and future planned increase of land-based wind turbines. However, the results also point to the triggers that cause the negative relation. The negative effects of seeing many turbines on a daily basis are tightly linked to having an on-land wind turbine in the direct view shed. Accordingly, many wind turbines in an area might not be a problem, as long as the number of respondents who have a direct view shed to turbines are minimised. In general, the cumulative effects of the total number of turbines on the individual attitude towards development pertains mainly to on-land turbines. As found in

Ladenburg & Möller (2010), attitude towards on-land wind turbines appears to be positively related to the number of wind turbines per wind farm. In constrast, in the offshore regions, people might actually prefer the turbines to be located in large wind farms as opposed to on-land turbines.

However, several studies did not find a significant effect of the number of wind turbines (see Krohn & Damborg, 1999; Ladenburg, 2010; Meyerhoff et al., 2010), but some of the applied parameters of prior experience might be too weak to establish an effect, such as the measure in Meyerhoff et al. (2010). In addition, the results in Krohn & Damborg (1999) might also be influenced by the fact that 58% of the respondents were co-owners of a wind turbine (see Ladenburg & Dahlgaard (2011) for a more detailed discussion of this issue). However, it is important to stress that the cumulative effects of existing wind turbine as described only have been statistically tested in several Danish studies, so it is difficult to generalise from the found cumulative effects. Since prior experience is difficult to capture, but highly relevant for spatial planning and management of rural areas, more research is needed.

4.2 View to wind turbines

Whether having an on-land or/and offshore wind turbine in the view shed from the permanent or summer residence or not, seems to have heterogeneous effects on the individual attitude. In general, having a wind turbine in the view shed seems to have some effects on the perception of wind farming. Having a view to on-land (and offshore) turbines or having a view to on-land turbines and seeing many turbines per day seems to reduce acceptability of a even stronger future expansion of on-land wind power systems (Ladenburg 2008; Ladenburg et al. 2011). On the other hand, having a view to on-land turbines can increase acceptability of offshore wind farms (Ladenburg, 2010). Interestingly, having a view to offshore wind farms appears not influence attitude towards offshore wind power (Ladenburg, 2008; 2009; 2010). This could point towards offshore wind power development becoming an increasing acceptable substitute for land-based wind power systems, if the future on-land development cannot be kept out of the view shed of peoples' residence. These first results from the offshore studies suggest that the present level of offshore wind power development does not seem to influence the attitude among the respondents who have offshore wind farms in the view. Accordingly, more offshore wind power development seems feasible from an attitude point of view.

4.3 Distance to turbines

The distance to wind turbines captures several dimensions of prior experience. If wind turbines are more common in the landscape, the distance captures the potential subjectivity to the impacts from the wind turbines. If people live close to a wind turbine, they might be more disturbed by visual intrusion, noise impact etc. compared to a respondent, who lives far from a wind turbine. However, if wind turbines are a relatively scarce commodity, such as the current offshore wind farms in Denmark, the distance captures a measure of the potential experience with a wind turbine, i.e. the further away people live from a wind turbine the lower is the probability that they have actually ever seen a wind turbine. Such effects could explain some of the observed distance effects in the reviewed studies.

Findings from the literature thus stress the role of distance, though the results are nonetheless ambiguous. In Warren et al. (2005) acceptance of on-land wind turbines decreases with distance. In Ladenburg & Möller (2010), this is also the case, however only to some extent. They find that respondents living within a 30-minute-drive from the nearest offshore wind farm are more negative. These results thus suggest an effect of living relatively close to wind farms. It is though important to note however that the measures of distance in the two studies are quite different. In Warren et al. (2005), the maximum distance the respondents live from the wind farm is 20 km. Compared to the analysis of distance in Ladenburg & Möller (2011), 20 km must be assumed to be within the 30 minutes of travelling time to the nearest offshore wind farm. Another distinct difference between the two studies is that Warren focuses on specific wind farms and does as such not control for the distances to other wind farms. In Ladenburg & Möller, the distance measure is to the nearest offshore wind farm, thus are not wind farm specific. This might also make it difficult to compare the results from the two studies. This stresses the difficulty to infer systematic relations in the effect of distance on attitude. This discussion should also be seen in the light of the results in Johansson & Laike (2007), who do not find an effect of distance.

4.4 Attributes of wind turbines

Conditional on having experience, it can be expected that the type of experience with the wind turbines has an impact on attitude. Ladenburg & Möller (2010) argue that the individual perception and actual exposure to wind turbines via the distance to the turbine(s), the size of the turbine(s) or the number of turbines in the vicinity might be very different between respondents. Accordingly, respondents living close to several large turbines are expected to have a completely different experience with wind turbines compared to respondents living near one single turbine, though both would state that they live close to a wind turbine. Some of these aspects were elaborated in the previous sections. However, to identify the links between the physical characteristics in terms of how different attributes of wind turbines and wind farms influence the individual attitude of a local resident remains to be a challenge. For instance, several studies have pointed out, that specific attributes of wind farms are preferred by the individual, such as locating wind farms offshore compared to on-land locations (Ek, 2005; McCartney, 2005), minimising the visual impacts from offshore wind farms etc. (Ladenburg & Dubgaard, 2007; Krueger et al., 2010 and number and size of wind turbines (Meyerhoff et al., 2010). However, the systematic influence of different wind farm characteristics on attitude has only been explored in a few studies to date. Though many studies analyse the effects from wind farms on the local community, to date only Ladenburg & Möller (2010) and Ladenburg (2009) have explicitly analysed if the variations in the wind farms affect the attitude of individuals within a local community differently.

Interestingly, there are systematic differences in offshore wind farm attributes. For instance, in Ladenburg (2009), differences in the visual impacts from offshore wind farms appear to have a significant impact on the attitude. If the offshore wind farms generate higher levels of visual impacts (the wind farm is located close to shore relative to the height of the wind turbines) more negative attitudes are generated. Apparently, differences in the size of the nearest offshore wind farm influence attitude, so that larger wind farms generate a more positive attitude. Interestingly, these results point towards that how (offshore) wind farms are planned and designed can have an positive influence on the acceptance of the wind farms. Hence, the results suggest that offshore wind farms should be located at relative large distances and should have rather more turbines in order to mitigate negative attitudes.

5. Conclusion

Prior experience with wind turbines is found to be a significant determinant of individual attitude towards wind farms in many studies. With the increasing level of wind power development on a global scale, this information can be of particular importance. The information entailed in the impacts from prior experience can thus serve as a guideline for policy planners and wind generation developers to increase the wind power capacity in an effective manner, so that opposition or negative attitudes towards wind power are minimised in future wind power landscapes.

Based on the significant prior experience effects, the review of the studies points towards that increasing number of turbines on-land can reduce the acceptance of future wind power development at even small additions to the current numbers of wind turbines. This is particularly evident if wind turbines cannot be kept out of the view shed from the individual residence. It could be shown that a solution to this increasing problem of local acceptance is apparently to move the future development offshore. Offshore, people seem to be less sensitive to view shed issues and the number of turbines. However, locating offshore wind farms too close to the shore might trigger even more negative attitudes.

6. References

Bishop, D. & Miller, D.R. (2007). Visual Assessment of Offshore Wind Turbines: The Influence of Distance, Contrast, Movement and Social Variables. *Renewable Energy*, Vol. 32, pp. 814-831.

AMR Interactive. (2010). Community attitudes to wind farms in NSW. Department of Environment, Climate Change and Water, New South Wales, Australia, available at http://www.environment.nsw.gov.au/resources/climatechange/10947WindFarms_Final.pdf

Ek, K. (2005). Public and Private Attitudes Towards Green Electricity: The Case of Swedish Wind Power. *Energy Policy*, Vol. 33, pp. 1677-89.

Graham, J.B.; Stephenson, J.R. & Smith, I.J. (2009). Public Perceptions of Wind Energy Developments: Case Studies from New Zealand. *Energy Policy*, Vol. 37, pp. 3348-3357.

Gross, C. (2007). Community Perspectives of Wind Energy in Australia: The Application of a Justice and Community Fairness Framework to Increase Social Acceptance. *Energy Policy*, Vol. 35, pp. 2727-2736.

Hagget, C. (2011). Understanding public responses to offshore wind power. *Energy Policy*, Vol. 39, pp.503-510,

Jobert, A.; Laborgne, P. & Mimler S. (2009). Local Acceptance of Wind Energy: Factors of Success Identified in French and German Case Studies. *Energy Policy*, Vol. 35, pp. 2751-2760.

Johansson, M. & Laike, T. (2007). Intention to Respond to Local Wind Turbines: the Role of Attitudes and Visual Perception. *Wind Energy*, Vol. 10, No. 5, pp. 435-451.

Jones, C.R. & Eiser, J.R. (2009). Identifying Predictors of Attitudes Towards Local Onshore Wind Development with Reference to an English Case Study. *Energy Policy*, Vol. 37, pp. 4604-4614.

Krohn, S. & Damborg, S. (1999). On Public Attitudes Towards Wind Power. *Renewable Energy*, Vol. 16, pp. 954-960.

Krueger, A.D.; Parsons G.E. & Firestone. J. (2010).Valuing the Visual Disamenity of Offshore Wind Power Projects at Varying Distances from the Shore. Working paper.

Kuehn, S. (2005). Sociological investigation of the reception of Horns Rev and Nysted offshore wind farms in the local communities. Report Econ analyse, available at http://www.hornsrev.dk/Miljoeforhold/miljoerapporter/Sociological_investigations_2003.pdf

Ladenburg, J. (2008). Attitudes Towards On-Land and Off-Shore Wind Power Development in Denmark: Choice of Development Strategy. *Renewable Energy*, Vol. 33, No. 1, pp. 111-118.

Ladenburg, J. (2009). Visual Impact Assessment of Offshore Wind Farms and Prior Experience. *Applied Energy*, Vol. 86. No. 3, pp. 380-387.

Ladenburg, J. (2010). Attitudes Towards Offshore Wind Farms - The Role of Beach Visits on the Demographic and Attitude Correlation. *Energy Policy*, Vol. 38, pp. 1297-1304.

Ladenburg, J. & Dahlgaard, J-O. (2011). Attitude threshold levels and cumulative effects of wind turbines in the local area. Working Paper, USAEE-IAEE WP 11-069.

Ladenburg, J. & Dubgaard, A. (2007). Willingness to pay for reduced visual disamenities from offshore wind farms in Denmark. *Energy Policy*, Vol. 35, pp. 4059–71.

Ladenburg, J. & Möller, B. (2010). Attitudes and acceptance of offshore wind farms in commission - The influence of travel distance and wind farm attributes. Manuscript, AKF, Danish Institute of Governmental Research.

Ladenburg, J.; Dahlgaard, J-O.; Hasler, B. & Termansen, M. (2010). Having a Wind Turbine in the View Shed or not: Cumulative Effects of Wind Turbines. Increasing the Wind Power Capacity in a Crowded Landscape - is less but larger better? Manuscript, AKF, Danish Institute of Governmental Research, 2010.

McCartney, A. (2005). The social value of seascapes in the Jurien Bay Marine Park: an assessment of positive and negative preferences for change. *Journal of Agricultural Economics*, Vol. 57, pp. 577–594.

Meyerhoff, J.; Ohl, C. & Hartje, W. (2010). Landscape Externalities from Onshore Wind Power. *Energy Policy*, Vol. 38, pp. 82-92.

Warren, C.R.; Lumsden, C.; O'Dowd, S. & Birnie, R.V. (2005). Green on Green: Public Perceptions of Wind Power in Scotland and Ireland. *Journal of Environmental Planning and Management*, Vol. 48, No. 6, pp. 853-875.

Part 2

Power Network Requirements

2

Wind Farms and Grid Codes

María Paz Comech, Miguel García-Gracia, Susana Martín Arroyo and Miguel Ángel Martínez Guillén
CIRCE-University of Zaragoza
Spain

1. Introduction

All customers connected to a public electricity network, whether generators or consumers, must comply with agreed technical requirements. Electric networks rely on generators to provide many of the control functions, and so the technical requirements for generators are unavoidably more complex than for demand customers. These technical requirements are termed 'Grid Codes'.

The technical requirements governing the relationship between generators and system operators need to be clearly defined. The introduction of renewable generation has often complicated this process significantly, as these generators have physical characteristics that are different from the directly connected synchronous generators used in large conventional power plants. In some countries, a specific grid code has been developed for wind farms, and in others the aim has been to define the requirements as far as possible in a way which is independent of the power plant technology.

The technical requirements within grid codes and related documents vary between electricity systems. However, for simplicity the typical requirements for generators can be grouped as follows:

- Tolerance - the range of conditions on the electricity system for which wind farms must continue to operate;
- Control of reactive power - often this includes requirements to contribute to voltage control on the network;
- Control of active power - often this includes requirements to contribute to frequency control on the network;
- Protective devices; and
- Power quality.

It is important to note that these requirements are often specified at the Point of Common Coupling (PCC) between the wind farm and the electricity network. In this case, the requirements are placed at wind farm level, and wind turbines may be adapted to meet these requirements. It is also possible for some requirements to be met by providing additional equipment, as for example for FACTS devices.

One of these new connection requirements regarding wind energy is fault ride-through capability. In the past, wind generators were not allowed to remain connected to the utility when voltage at the PCC fell below 85 %, forcing their disconnection even when the fault happened far from the wind farm (Jauch et al, 2007; Rodriguez et al, 2002). That is the reason why, in grids with significant wind energy penetration, the voltage dip and the subsequent wind farm disconnections would create an important stability problem.

Therefore, it is important to check the compliance with Grid Codes. The Spanish Wind Energy Association has developed the document "Procedure for Verification Validation and Certification of the Requirements of the OP 12.3 on the Response of Wind Farms in the Event of Voltage Dips (PVVC) (AEE, 2007), and the German Fördergesellschaft Windenergie und andere Erneuerbare Energien the document "Technical Guidelines for Power Generating Units. Part 8. Certification of the electrical characteristics of power generating units and systems in the medium., high- and highest-voltage grids"(FGW-TG8) (FGW, 2009) that describes the procedures to certify wind power installations according their corresponding Grid Codes.

The Compliance with Grid Codes can be checked by means of in-field test or by simulation of validated models. This chapter describes the procedure to verify wind installations according PVVC and FGW-TG8. Section 2 lists the most outstanding international Grid Codes, section 3 describes the fault ride through solutions of the different wind turbine types. Section 4 describes the fault ride through certification procedure, section 5 the voltage dip test, section 6 the model validation according to PVVC and FGW-TG8. Section 7 the wind farm verification according to PVVC.

2. International grid code requirements

Wind farms should contribute to power system control (voltage and frequency) and also to the electricity network recover in case of networks faults such as voltage dips or swells.

In the most cases a wind turbine should work with a power factor of 0.90 lagging to 0.95 leading and the frequency should situate within the range from 47.5 Hz to 52 Hz.

The most outstanding international Grid Codes are the following:

- USA FERC: *"Interconnection for Wind Energy"* 18 CFR Part 35 (Docket No. RM05-4-001; Order No. 661-A), Issued December 12, 2005 and *"Interconnection Requirements for a Wind Generating Plant"*, Appendix G to the LGIA.
- Germany - E.ON Netz GmbH: *"Grid Code - High and extra high voltage"*, Status: 1.April 2006.
- China - CEPRI: *"Technical Rule for Connecting Wind Farm to Power System"*, December, 2005.
- Spain - REE - P.O. 12.3: Resolución de 4 de octubre de 2006, de la Secretaría General de Energía por la que se aprueba el procedimiento de operación 12.3 *"Requisitos de respuesta frente a huecos de tensión de las instalaciones eólicas"*. Publicación en BOE núm. 254 de fecha 24 Octubre 2006.
- India - ISTS: *"Indian Electricity Grid Code (IEGC)"*, April, 2006 and *"Draft Report on Indian Wind Grid Code"*, July, 2009.
- France: *"Décret no 2008-386 du 23 avril 2008 relatif aux prescriptions techniques générales de conception et de fonctionnement pour le raccordement d'installations de production aux réseaux publics d'électricité"*, April, 2008.
- Italy: *"CEI 11-32; V1 Impianti di produzione eolica"*, December, 2006.
- Great Britain - National Grid Electricity Transmission plc: *"The Grid Code"*, Issue 4 Revision 3, 6th September 2010.
- Denmark - ELKRAFT SYSTEM and ELTRA: *"Wind Turbines Connected to Grids with Voltages above 100 kV - Technical regulations for the properties and the regulation of wind turbines"*, Regulation TF 3.2.5, December 3, 2004.

- Portugal - REN: Portaria n.° 596/2010 de 30 de Julho
- Canada - AESO: *"Wind Power Facility - Technical Requirements"*, Revision 0, November, 15 2004.
- Australia - AEMC: *"National Electricity Rules (NER)"*, Version 39, 16 September 2010
- Ireland - EIRGRID: *"WFPS1- Controllable Wind Farm Power Station Grid Code Provisions"*, EirGrid Grid Code, Version 3.4, October 16th 2009.

Fault ride through requirements are described by a voltage vs. time characteristic, denoting the minimum required immunity of the wind power station. The fault ride through requirements also include fast active and reactive power restoration to the prefault values, after the system voltage returns to normal operation levels. Some codes impose increased reactive power generation by the wind turbines during the disturbance, in order to provide voltage support, a requirement that resembles the behaviour of conventional synchronous generators in over-excited operation.

Fig. 1 presents in the same graph the fault ride through requirements from the different Grid Codes. These requirements depend on the specific characteristics of each power system and the protection employed and they deviate significantly from each other.

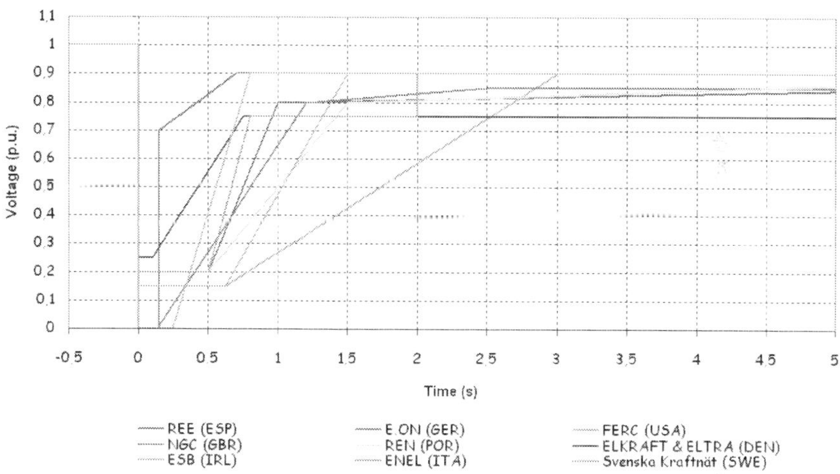

Fig. 1. Fault ride through requirements.

3. Wind turbine fault-ride through

As it has been said, one of the main problems for power quality are voltage dips. Due to high renewable penetration level in transmission system, Transmission System Operators (TSO) demand to this sort of energy source support voltage under voltage sags. This obligation has provoked a huge investment in devices to support wind systems during voltage dips.

Fig. 2 shows the three main technologies in the wind turbine industry. Their behaviour is different in continuous operation and during voltage dips.

Fig. 2a shows the fixed-speed wind turbine with asynchronous squirrel cage induction generator (SCIG) directly connected to the grid via transformer. Fig. 2b represents the limited variable speed wind turbine with a wound rotor induction generator and partial scale frequency converter on the rotor circuit known as doubly fed induction generator (DFIG). Fig. 2c shows the full variable speed wind turbine, with the generator connected to the grid through a full-scale frequency converter.

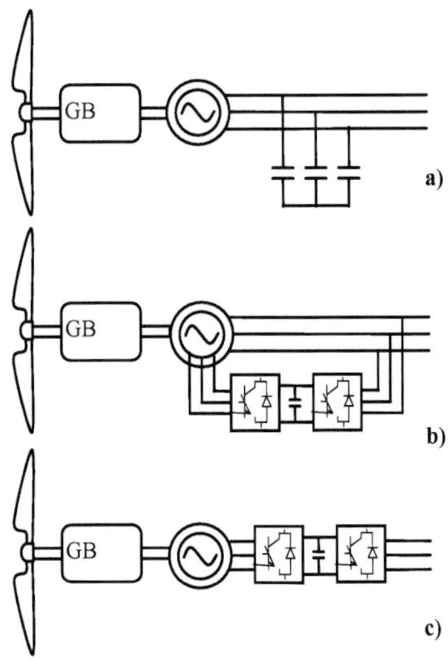

Fig. 2. Wind turbine technologies.

DFIG stator is connected directly to the network but its rotor is connected to the network by means of a power converter which performs the active and reactive power control. A voltage dip will cause large currents in the rotor of the DFIG to which the power electronic converter is connected, and a high rotor voltage will be needed to control the rotor current. When this required voltage exceeds the maximum voltage of the converter, it is not possible any longer to control the current desired (Morren, de Haan, 2007). This implies that large current can flow, which can destroy the converter.

In order to avoid breakdown of the converter switches, new DFIG wind turbines are provided with a system called crowbar connected to the rotor circuit. When the rotor currents become too high, the converter is disconnected and the high currents do not flow through the converter but rather into the crowbar resistances. The generator then operates as an induction machine with a high rotor resistance. When the dip lasts longer than a few hundreds of milliseconds ($T_{max\_crowbar}$), the wind turbine can even support the grid during the dip (Morren, de Haan, 2007; López et al, 2009).

The full converted wind turbine is connected to the network through a converter; and therefore the converter controls the wind turbine during de dip in order to fulfill the Grid Code Requirements.

SCIG are used as fixed speed wind generator due to its superior characteristics such as brushless and rugged construction, low cost, maintenance free, and operational simplicity. However it requires large reactive power to recover the airgap flux when a short circuit occurs in the power system, unless otherwise the induction generator becomes unstable due to the large difference between electromagnetic and mechanical torques, and then it requires to be disconnected from the power system (Muyeen et al, 2009; Muyeen & Takahashi, 2010). Next section describes different solutions to support the transient behaviour of SCIG and old DFIG wind turbines that do not fulfill fault ride through requirements.

3.1 Fault ride through solutions

Nowadays, the rapid development of power electronics has made that the old devices for controlling voltage based on capacitors and reactors have been replaced by Flexible AC Transmission Systems (FACTS).

New wind turbines have integrated different systems to withstand voltage dips; however the old wind turbines have to install different FACTS to overcome dips. The main solutions are installed either in each turbine or in the point of common coupling.

The FACTS used in wind systems can be divided into three categories depending on their connection (Amaris, 2007; Hingorain, 1999):

- Series device, for example the Dynamic Voltage Restorer (DVR)
- Shunt device, such as Static Voltage Compensator (SVC) and Static Compensator (STATCOM).
- Series-shunt device. They are a combination of a series and a parallel FACTS. In wind system Unified Power-Quality Conditioner (UPQC) are used.

Next, these systems are explained.

3.1.1 Static Voltage Compensator (SVC)

Static Voltage Compensator is a shunt-connected var generator o absorber whose output is adjusted to exchange capacitive or inductive current. Fig. 3 shows the connection of SVC. It is usually connected between the utility and the generator. SVC can provide reactive power, from 0 to 1 p.u. depending on voltage (Fig. 3). These devices use electronic switches as thyristor, which can open or close in few milliseconds. SVC is considered by some as a lower cost alternative to STATCOM, although this may not be the case if the comparison is made based on the required performance and not just in the MVA size, because for the same contingency and the same system, the required SVC ratings is generally larger than required STATCOM (Hingorain, 1999, Molinas et al, 2008).

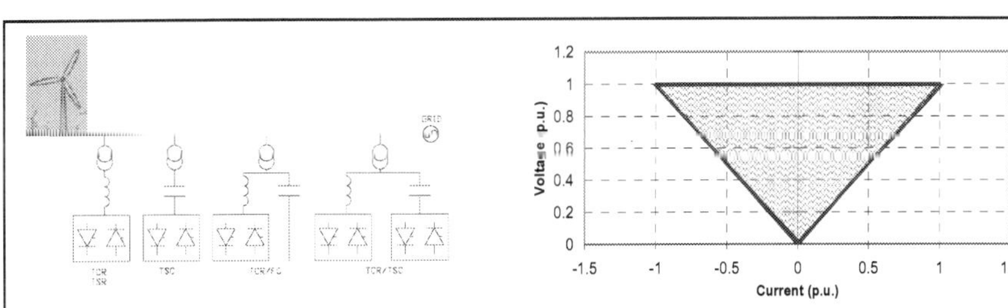

Fig. 3. Different topologies of SVC and V-I characteristic.

3.1.2 Static Synchronous Compensator (STATCOM)

Static Synchronous Compensator is a voltage source converter which can inject or absorb reactive current in an AC system, modifying the power flow. STATCOM can provide

reactive power independently of the voltage, as shown the voltage-current characteristic in Fig. 4. It comprises a converter, connected in parallel between utility and the generator, and a DC current stage as it is shown in Fig. 4.
STATCOM is the evolution of SVC, but STATCOM have continuous control and can compensate both power factor and voltage simultaneously. Other advantage of STATCOM is its dynamic capacity getting small response times.

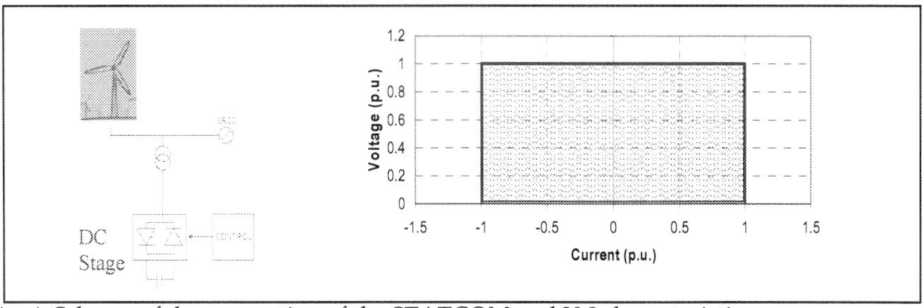

Fig. 4. Scheme of the connection of the STATCOM and V-I characteristic.

3.1.3 Dynamic Voltage Restorer (DVR)

Dynamic Voltage Restorer is a series compensator, which works inserting a voltage of magnitude and frequency necessary. Fig. 5 shows the scheme of this FACTS.
DVR consists of a medium voltage switchgear, a coupling transformer, filters, rectifier, inverter, and energy source (e.g. storage capacitor bank) and control and protection system. DVR can inject or absorb real and reactive power independently by an external storage system without reactors and capacitors (Wizmar & Mohd, 2006).
If the storage system is a capacitor bank, during normal operation it will be charging, and when a swell or voltage sag is detected this capacitor will discharge to maintain load voltage supply injecting or absorbing reactive power.

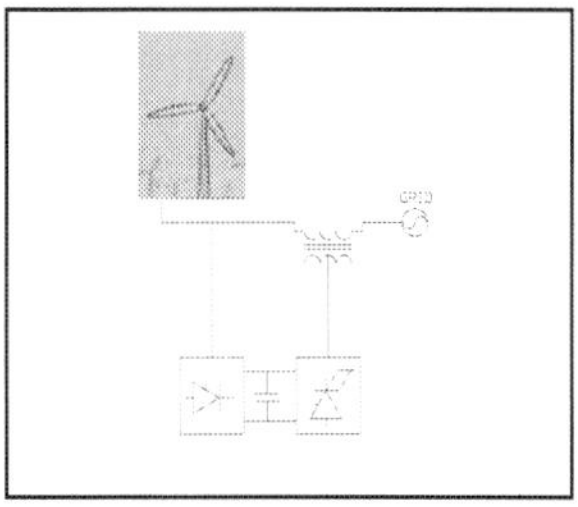
Fig. 5. Scheme of Dynamic Voltage Restorer.

3.1.4 Unified Power Quality Conditioner (UPQC)

Unified Power Quality Conditioner is a combination of a series and a shunt FACTS. Its target is to improve power quality compensating voltage flicker, unbalance, negative-sequence current and harmonics. Fig. 6 shows the scheme of connection of UPQC.
UPQC (Khadkikar et al, 2004) comprises two voltage source inverters connected back to back and sharing a dc link. The shunt inverter helps in compensating load harmonic current

and maintains dc voltage at constant level. The second inverter is connected in series by using a series transformer and helps in maintaining the load voltage sinusoidal and compensate voltage dips and swells.
Control system of UPQC is formed by the positive sequence detector, the series inverter control and the shunt inverter control.

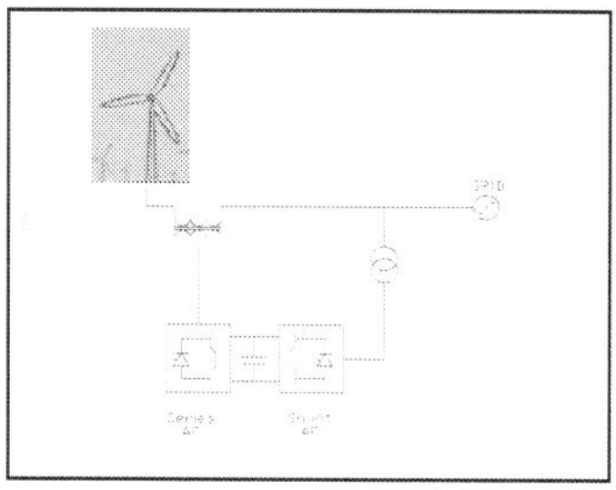

Fig. 6. Scheme of Unified Power-Quality Conditioner.

4. Fault ride through certification procedure for power generating units

Once the requirements for wind power system have been established, another important point is how wind turbine manufacturers and wind park operators can prove the fulfilment of Grid Codes. The Spanish Wind Energy Association (AEE) has developed the document "Procedure for Verification Validation and Certification of the Requirements of the OP 12.3 on the Response of Wind Farms in the Event of Voltage Dips" (PVVC), and the German Fördergesellschaft Windenergie und andere Erneuerbare Energien (FGW) the document "Technical Guidelines for Power Generating Units. Part 8. Certification of the electrical characteristics of power generating units and systems in the medium, high- and highest-voltage grids" that describes the procedures to certify wind power installations according their corresponding Grid Codes.
This section describes the steps to fulfil certificate wind systems by these two procedures.

4.1 PVVC procedure

The PVVC define two possible processes to verify the conformity with the response requirements established in OP 12.3:

- The General Verification Process
- The Particular Verification Process

The General Verification Process consists of verifying that the wind farm does not disconnect and that the requirements stated on the OP 12.3 are met by means of:

- Wind turbine and/or FACTS test
- Wind turbine and/or FACTS validation
- Wind farm simulation

Then three processes must be followed to verify an installation by the General Verification Process and three reports are needed. Next figures show a scheme of these three processes and the three reports obtained. Fig. 7 shows the scheme of the field test process, Fig. 8 the model validation process and Fig. 9 the verification process.

Fig. 7. Field test process.

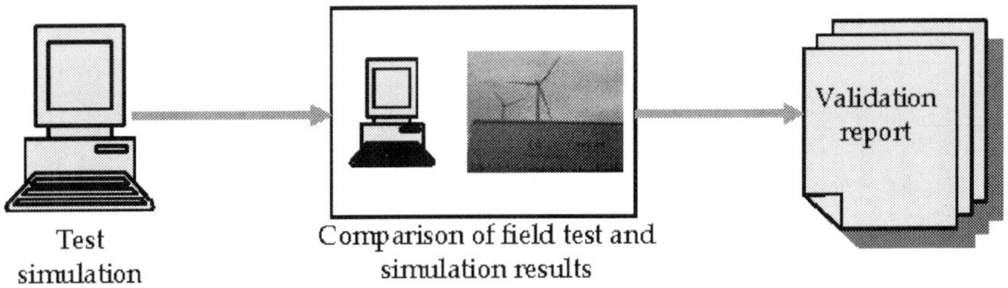

Fig. 8. Validation process.

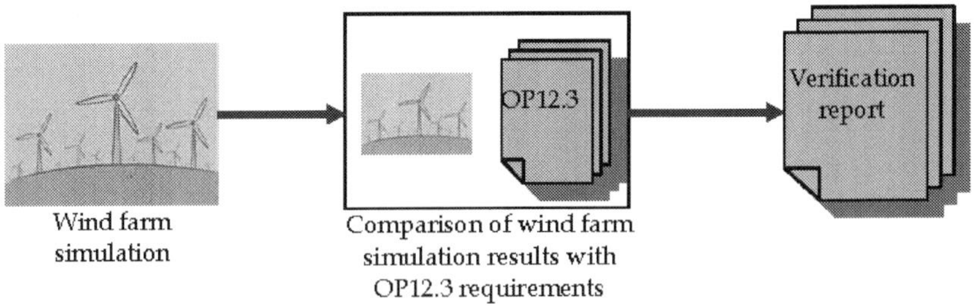

Fig. 9. Verification process.

The particular verification process obtains the direct wind farm verification by testing the dynamic elements of the wind farm. In this case, only the process shown in Fig. 7 must be performed. Model validation and wind farm simulation are not needed. In this case, the conditions of the field test will be harder than those of the general verification process.

The particular verification process is faster and cheaper than the general verification process. Therefore, wind turbine manufacturer and wind farm operators would prefer this process if the wind turbine or the system wind turbine + FACTS can be tested and can ride through the voltage dip test defined in the Particular Verification Process. General Verification Process is necessary in those wind farms whose wind turbines can not ride through the voltage dip defined in the particular process and a compensating system is installed on the wind farm substation to fulfil the OP 12.3 requirements.

4.2 FGW-TG8 procedure

The FGW-TG8 defines two processes depending on the date of commission of the installation that is going to be certificate. If the installation has been commissioned after 01.01.2009 must follow the process for "new generating units". If the installation has been commissioned after 31.12.2001 and before 01.01.2009 the certification must follow the process for "old systems".

To certify "new generating units" the applicant must provide:

- Verification of type testing according to FGW-TG3 (FGW, 2009).
- A comprehensive computer based model of the power generating unit, which may be encapsulated as a black box model. This model needs to be suitable to represent the measuring situation of the type tests in accordance with FGW-TG3 (FGW, 2009).
- An open, where necessary simplified, model of the power generating unit. This open model must allow the certifier to follow the logical links between control loops in the relevant system controls. The degree of detail of the open model must be clarified in advance between the certification authority and the manufacturer. In some cases it may be sufficient to present block diagrams. It is necessary to comprehensively describe fault detection for verification of performance in a fault situation.

To certify "old systems" the applicant must provide Verification of type testing according to FGW-TG3. Furthermore the document must contain the specification of the original power generating unit and the specifications on the refitted power generating unit. Model validation does not form part of this procedure.

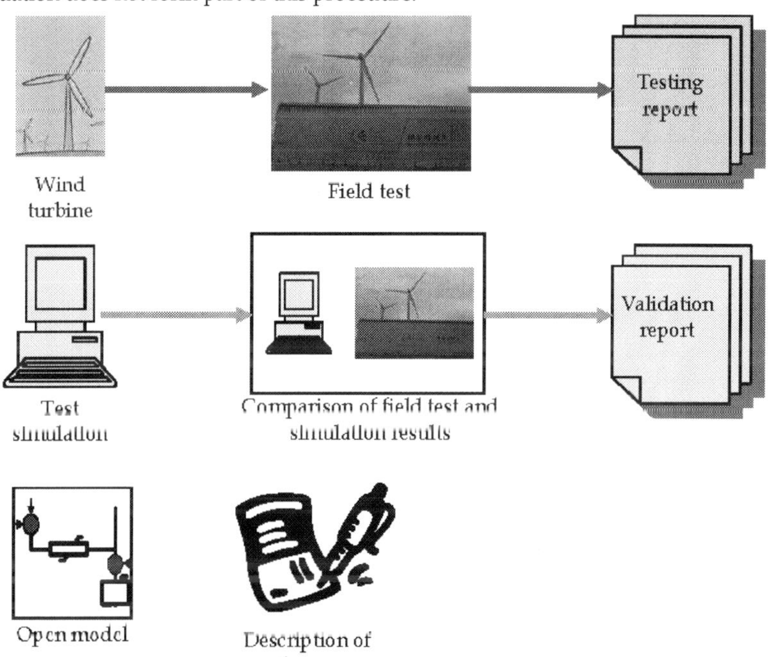

Fig. 10. Process of new unit certification.

Fig. 11. Process of old unit certification.

5. Voltage dip test

In order to test the behaviour of the turbine when a voltage dip occurs and the compliance with Grid Codes, a device able to generate voltage dips is required. This device must create a voltage variation according to the regulations of the different countries in order to check that the tested wind turbine fulfils the established requirements, such as voltage ride-through, short circuit contribution and power factor.

5.1 Voltage dip generator

Voltage dip generators are based on the use of two impedances, as it is shown in Fig. 12 (Niiranen, 2005, 2006; Gamesa eólica, 2006; Gamesa innovation and technology, 2006). The parallel impedance enables the generation of the fault while the series impedance immunizes the grid from the dip and the test can be performed without affecting other systems connected to it.

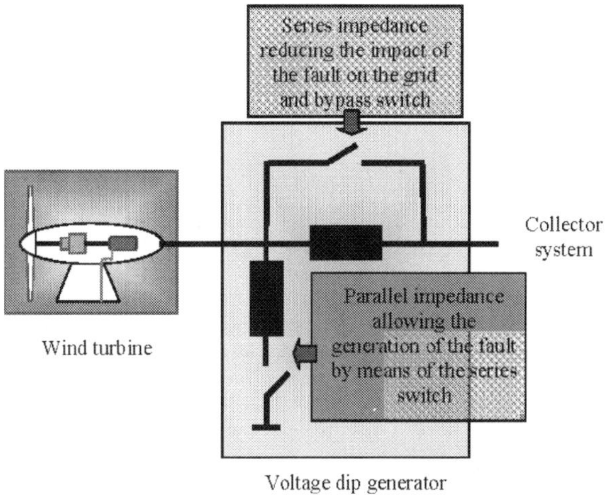

Fig. 12. Dip generator scheme and its position with respect to the windmill and the wind farm.

5.1.1 20 kV 5 MW Voltage dip generator

This section describes the design of a 20 kV, 5 MW voltage dip generator. It is installed in a trailer, so it is able to move to the wind turbine location (García-Gracia et al, 2009).

Fig. 14 shows a scheme of this voltage dip generator. It is based on an inductive divider comprised of a series and a parallel branch, and its main components are a three-phase series impedance (4) at the system input, a parallel tap transformer (7) and a three-phase impedance (11) grounded through a control switch in the secondary of the transformer. This impedance allows the adjustment of the dip depth to the desired value, along with the regulation of the transformer, because the impedance (11) connected to winding 2 is referred to winding 1 by multiplying by the square of the turns ratio. Switches (5) and (9) make possible the generation of a 100% depth voltage dip.

Fig. 13. Picture of the 5 MW test system.

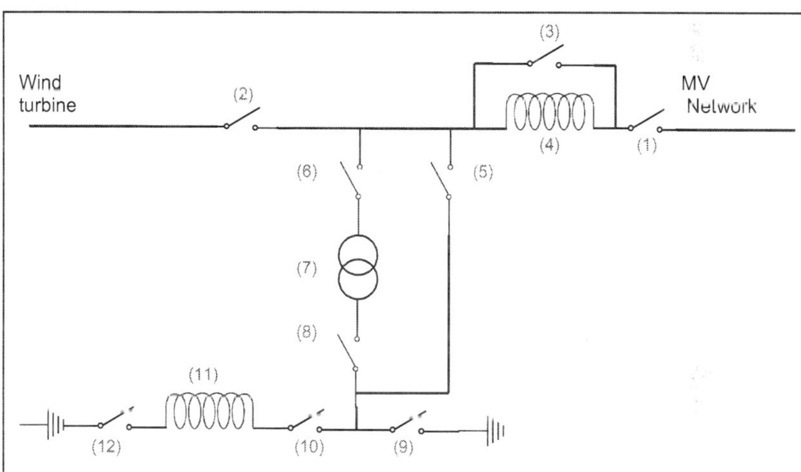

Fig. 14. Scheme of the voltage dip generator.

5.2 Voltage dip test procedure

The system described includes some other control elements in order to perform the voltage dip generation, which takes place as follows.

Having the by-pass switch (3) on allows the direct connection between the utility and the generating system (i.e. wind system), eliminating the effect of the insertion of the voltage dip generator.

Once this switch is open, the generator is connected to the grid through the series inductances (4), and the switch (6) connecting the parallel branch can be closed, in order to connect the primary of the transformer (7), which at this point is in no-load operation.
Next, the dip generation switch (8) is closed, connecting the secondary of the transformer to the impedances (11) or to the short circuit (9) to achieve a deeper voltage dip. Timing the operation of these switches, the desired dip duration is set. As mentioned before, a 100% voltage dip can be achieved closing switches (5) and (9) after switch (3) has been open.
The impedance banks (11) have single-phase switches (10) to have the possibility of performing single-phase, two-phase and three-phase tests.

5.2.1 Wind turbine test according to the Spanish PVVC

The Spanish PVVC distinguish between two different type tests:

- Test for validating the simulation model (General Verification Process)
- Test for direct observance of the OP 12.3 (Particular Verification Process)

For both cases, the wind turbine should be tested for the following operation points:

	Registered Active Power	Power Factor
Partial load	10% - 30% Prated	0.9 inductive – 0.95 capacitive
Full load	> 80% Prated	0.9 inductive – 0.95 capacitive

Table 1. Operation points prior to test.

The depth of the voltage dip must be independent of the wind turbine tested. Therefore, a no- load test must be performed before the connection of the wind turbine. Thus the series inductances (4), the transformer taps (7) and the impedances (11) are adjusted with the switch (2) open.
Table 2 shows the residual voltage, the duration of the voltage dips, and the allowed tolerances of the tests for direct observance of the OP 12.3 (Particular Verification Process).

Dip	Residual dip voltage (Ures)	Voltage tolerance (Utol)	Dip duration (ms)	Time tolerance (Ttol) (ms)
Three phase	≤(20%+Utol)	+ 3%	≥ (500-Ttol)	50
Isolated two phase	≤(60%+Utol)	+ 10%	≥ (500-Ttol)	50

Table 2. Voltage dip properties in the no-load test for the Particular Verification Process.

If the objective of the test is the validation of simulation models (General Verification Process), the minimum voltage registered during the no load test of the faulted phases must be less than 90%.
Before the wind turbine test, it must be checked that the short circuit power in the test point is greater than 5 times the generator rated power. This condition is fulfilled by adjusting (4).
Once the voltage dip generator has been adjusted; the test can be performed by closing the switch (2) of the Fig. 14. The four test categories shown in Table 3 must be carried out. Therefore, the power generated by the wind turbine must be measured before the voltage dip, to check the operating point. As the operating point depends on the wind speed, it is possible that the generated power does not match with one of the operating points shown in Table 1. In this case, the laboratory has to wait for the needed weather conditions to perform the test of each operating condition.

Category	Operating point	Dip type
1	Partial load	Three phase
2	Full load	Three phase
3	Partial load	Isolated two phase
4	Full load	Isolated two phase

Table 3. Test categories.

Fig. 15 and Fig. 16 show the measured voltages during a three-phase and a two-phase voltage dip respectively.

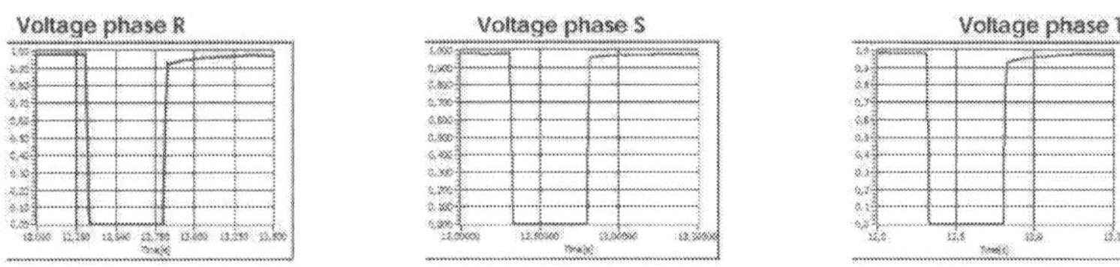

Fig. 15. Three-phase voltage dip: Depth 100%; Duration 510 ms.

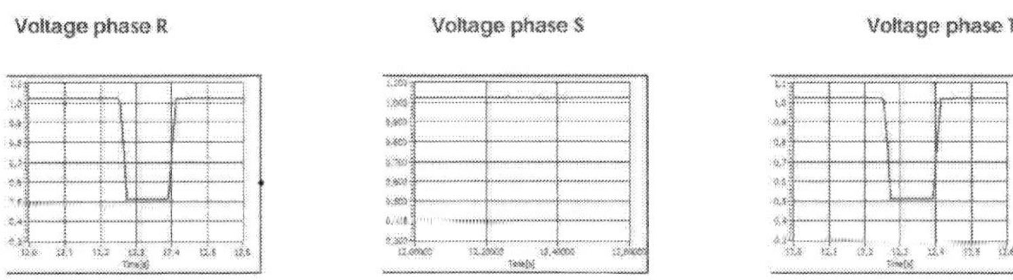

Fig. 16. Two-phase voltage dip: Depth 50%; Duration 150 ms.

To guarantee the continuity of supply, the wind turbine will be undergone to three consecutive tests. If the wind turbine disconnects during this test sequence, four consecutive tests will be performed. If in this new sequence, the wind turbine disconnects, the test will be considered invalid.
To verify wind systems by applying the Particular Verification Process, the power and energy registered must fulfill the requirements shown in Table 4 and Table 5.

Three phase faults	OP 12.3 requirements
ZONE A	
Net consumption Q < 15% Pn (20 ms)	-0.15 p.u.
ZONE B	
Net consumption P < 10% Pn (20 ms)	-0.1 p.u.
Net consumption Q < 5% Pn (20 ms)	-0.05 p.u.
Average I_r/I_{tot}	0.9 p.u.
Extended ZONE C	
Net consumption $I_r < 1.5\ I_n$ (20 ms)	-1.5 p.u.

Table 4. Power and energy requirements for three phase voltage dips in the Particular Verification Process.

Two phase faults	OP 12.3 requirements
ZONE B	
Net consumption E_r < 40% Pn * 100 ms	-40 ms·p.u.
Net consumption Q < 40% Pn (20 ms)	-0.4 p.u.
Net consumption E_a < 45% Pn * 100 ms	-45·ms p.u.
Net consumption P < 30% Pn (20 ms)	-0.3 p.u.

Table 5. Power and energy requirements for isolated two phase voltage dips in the Particular Verification Process.

Where the zones A, B and C are defined in Fig. 17.

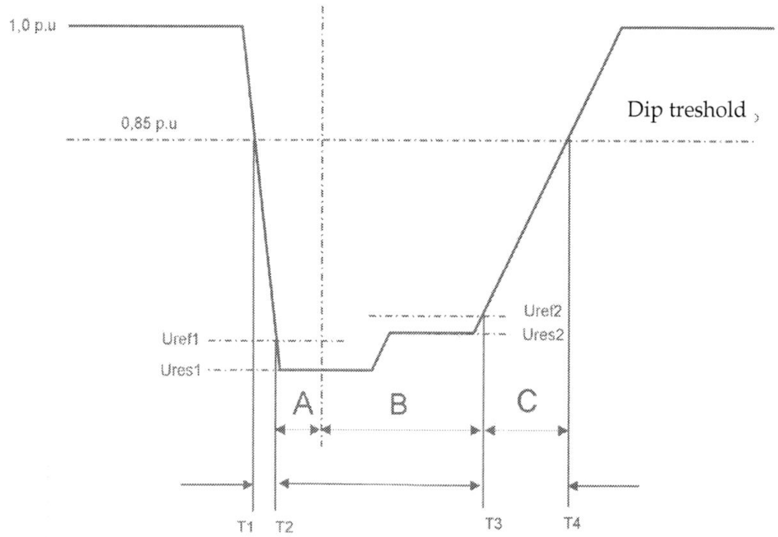

Fig. 17. Classification of the voltage dip in the field test.

5.2.2 Wind turbine test according to the German FGW-TG3

The on-site test should serve the following objectives:

- Validation of the system
- Test the control system and the auxiliary units

For both cases, the wind turbine should be tested for the following operation points:

	Registered Active Power
Partial load	10% - 30% Prated
Full load	> 90% Prated

Table 6. Operation points prior to test.

In this case, the voltage dip generator must have an X/R ratio of at least 3, and the symmetrical fault level on the transformer's high voltage side must be at least 3·Prated.

The voltage dip generator must be configured in no-load test to obtain the three phase and two phase voltage dips with the different depths shown in Table 7 for directly synchronous generators and Table 8 for the other types, as in the procedure for test according to the Spanish PVVC. Therefore, in the system shown in the Fig. 14, the series inductances (4), the transformer taps (7) and the impedances (11) adjusted with the switch (2) open.

Test number	Ratio of fault voltage to initial voltage (U/U0)	Fault duration (ms)
1	0.05	150
2	0.20-0.25	150
3	0.45-0.55	150
4	0.70-0.80	700

Table 7. Voltage drop test for directly coupled synchronous generators.

Test number	Ratio of fault voltage to initial voltage (U/U0)	Fault duration (ms)
1	0.05	150
2	0.20-0.25	550
3	0.45-0.55	950
4	0.70-0.80	1400

Table 8. Voltage drop test for all the other types of generators.

For three phase voltage dips in accordance with test 3 and 4, minimum proportionality constant (K-factor) is two. This factor is defined in (SDLWindV, 2009) by:

$$\frac{\Delta I_B}{I_N} = K \cdot \frac{\Delta U_r}{U_N} \tag{1}$$

Where I_B is the reactive current, ΔI_B is the reactive current deviation and ΔU_r is the relevant voltage deviation and is calculated as:

$$\Delta U_r = \Delta U + U_t \tag{2}$$

Where ΔU is the voltage deviation and U_t the dead band, that must be kept at a constant maximum of 10% U_N during each test.

6. Model validation

The Spanish PVVC and the German FGW-TG4 (FGW, 2009) give the procedures to validate wind turbine systems by comparing the results obtained by simulation and that obtained from on-site test. PVVC and FGW-TG4 gives the maximum deviation and the specific time intervals for the comparison of the results. The Spanish PVVC establishes a time window of 1 s with 100 ms before the voltage dip, and the German FGW-TG4, 500 ms before the voltage dip and 2 s after the voltage recovery. Fig. 18 shows the different time windows established in each document. It is important to point out that the time window from the PVVC is fixed and does not depend on the voltage dip duration whereas the FGW-TG4 depends on it.

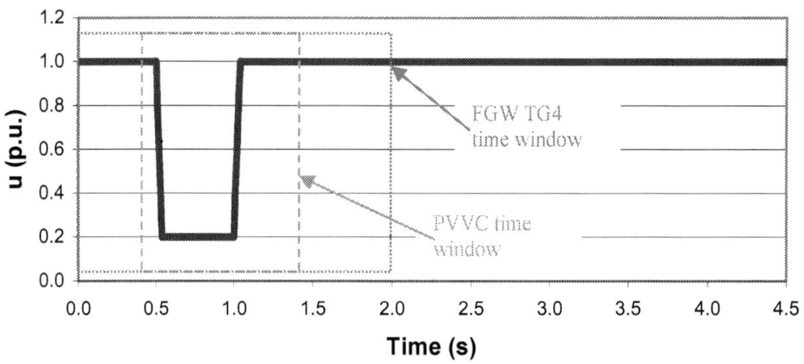

Fig. 18. Time window established in the German FGW-TG4 and the Spanish PVVC.

Respect the maximum deviation, in the Spanish PVVC it is constant and equal to 10% in the time frame, and the German FGW-TG4 establishes these values:

	Deviation F1	Deviation F2	Deviation F3	Total Deviation FG
Active Power $\Delta P/Pn$, Reactive Power $\Delta Q/Pn$	0.07	0.20	0.10	0.15
Reactive current $\Delta Ib/Ir$	0.10	0.20	0.15	0.15

Table 9. Maximum deviation in different stages of voltage dip.

Where F1 is the deviation of the mean of steady state areas, F2 the deviation of the mean of transient areas, F3 the highest deviation in steady state areas and FG the mean of weighted deviations for P, Q and Ib.

Next the validation process followed for a wind turbine generator from in-field testing results according to the Spanish PVVC.

6.1 Voltage dip generator model

In PVVC the system shown in Fig. 19 is proposed. In this system, the voltage measured in the field test is introduced in the simulation and reproduced by a voltage source. Thus, the wind turbine model is subjected to the same voltage than the wind turbine during the field test and only the active and reactive power must be compared to validate the model.

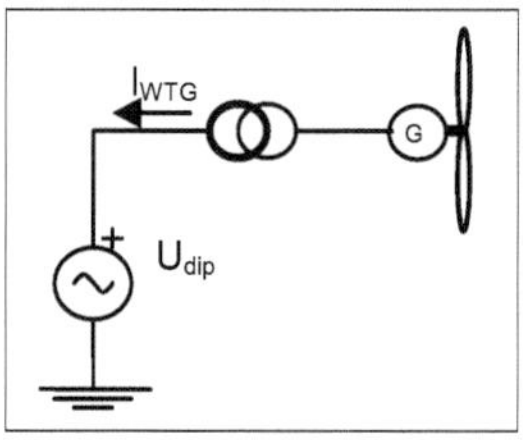

Fig. 19. Voltage dip generator representation in validation simulation.

6.2 Methodology for calculating power

The PVVC explains the following method to calculating power from the test and simulation results.

Using the N samples of the instantaneous values of phase voltage (u(n)) and the phase current (i(n)) the fundamental harmonic can be obtained using the following expressions:

$$\underline{U}_1 = \frac{\sqrt{2}}{N} \cdot \sum_{n=0}^{N-1} u(n) \cdot e^{-j\left(\frac{2\pi n}{N}\right)} \tag{5}$$

$$\underline{I}_1 = \frac{\sqrt{2}}{N} \cdot \sum_{n=0}^{N-1} i(n) \cdot e^{-j\left(\frac{2\pi n}{N}\right)} \tag{6}$$

To calculate the active and reactive power, only the positive sequence component of the voltage and current are used:

$$\underline{U}^{+} = \frac{1}{3}\left(\underline{U}_{1A} + \underline{U}_{1B} \cdot e^{+j\frac{2\pi}{3}} + \underline{U}_{1C} \cdot e^{-j\frac{2\pi}{3}}\right) \tag{7}$$

$$\underline{I}^{+} = \frac{1}{3}\left(\underline{I}_{1A} + \underline{I}_{1B} \cdot e^{+j\frac{2\pi}{3}} + \underline{I}_{1C} \cdot e^{-j\frac{2\pi}{3}}\right) \tag{8}$$

The three-phase active and reactive power expressions are obtained from the positive sequence component of the voltage and current as:

$$P = 3 \cdot U^{+} \cdot I^{+} \cdot \cos(\varphi) \tag{9}$$

$$Q = 3 \cdot U^{+} \cdot I^{+} \cdot sen(\varphi) \tag{10}$$

6.3 Model validation

This section describes the model validation process followed for the developed model. Only the three-phase voltage dip for the full load category is shown, the process for the rest of the categories would be the same.

The next figure shows the voltage evolution during the field test and the simulation in phase A. In the simulation, the voltage is introduced by means of a voltage source that reproduces the voltage during the field test. Therefore, there are no significant differences between test and simulation. Voltage in phase B and C are similar to voltage in phase A. In the figure, the blue line represents the voltage obtained during the field test; the red line has been obtained by simulation and the green line the maximum deviation considered in the Spanish PVVC (10%).

Table 10 shows that the model is validated in this category (full load, three phase voltage dip) because the number of the samples with error less than the maximum allowable error for the active and the reactive power are greater than 85%. Fig. 21 shows the comparison of the active power results and Fig. 22 the comparison of the reactive power results. In both figures, the blue line represents the results obtained during the field test; the red line has been obtained by simulation and the green line the maximum deviation considered (10%).

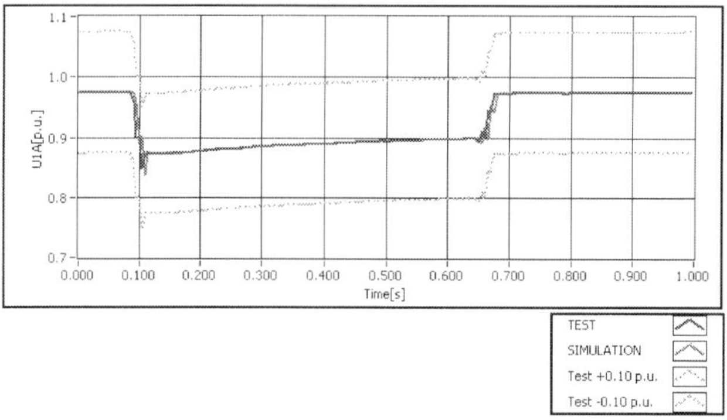

Fig. 20. Voltage evolution during the field test and the simulation in phase A.

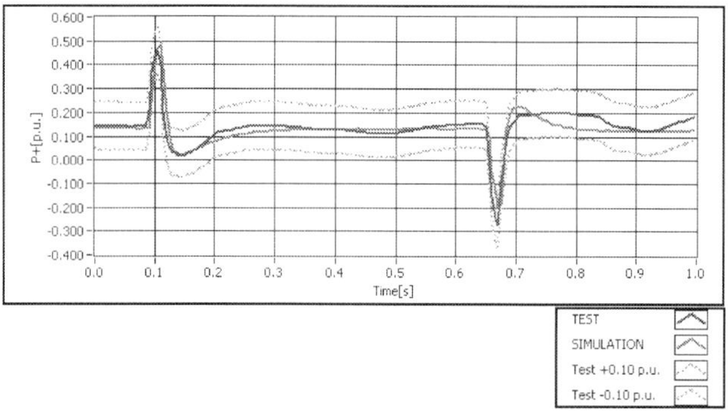

Fig. 21. Comparison of the active power during field test and simulation.

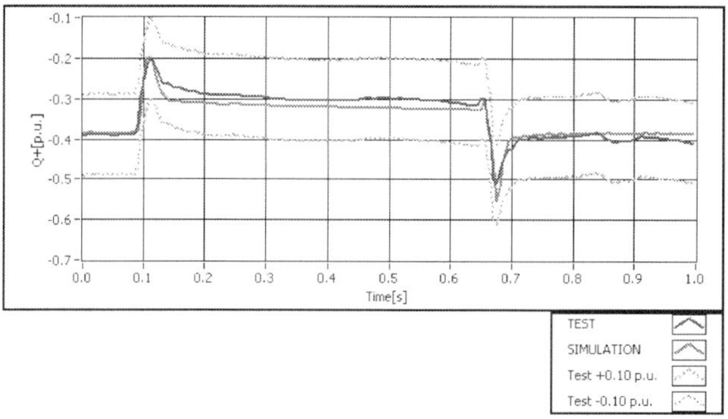

Fig. 22. Comparison of the reactive power during field test and simulation.

¿Is the model validated?	Yes
P samples with error < 0.1 p.u.	97.50
Q samples with error < 0.1 p.u.	100.00

Table 10. Validation results for the example.

7. Wind farm verification

As it has been shown in section 4.1, if the General Verification Process of the PVVC is followed, a simulation study must be performed. The simulation tool used to verify wind installation according to PVVC must permit to model the electrical system components per phase, because balanced and unbalanced perturbances must be analyzed.
The simulated model to verify the installation must take into account the different components of the real system, that is: the wind farm, FACTS and reactive compensating systems, the step-up transformer, the connection line and a equivalent network defined in PVVC. Fig. 23 shows the one line diagram of the network to be simulated.

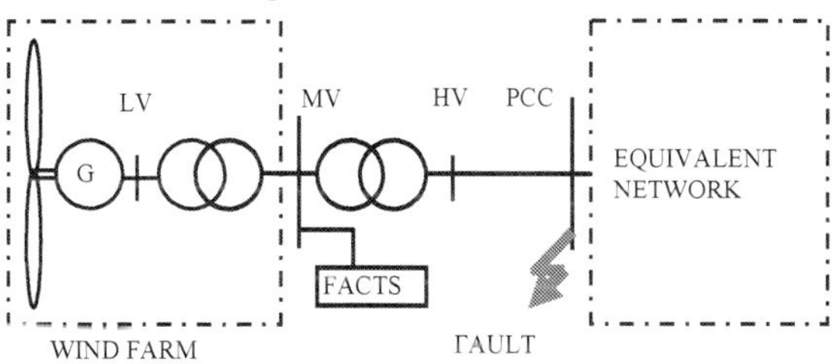

Fig. 23. One line diagram of the wind installation network.

The PVVC establishes the external network model equivalent. This equivalent network reproduces the typical voltage dip profile in the Spanish electrical system, that is a sudden increase in the moment of the clearance and a slower recovery afterwards. The profile for three phase voltage dips is shown in Fig. 24.

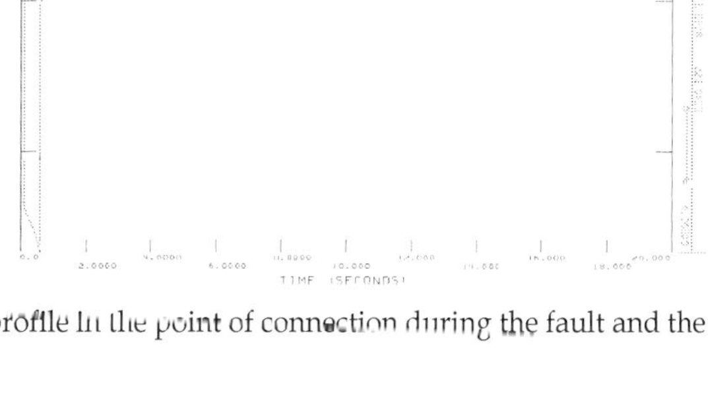

Fig. 24. Voltage profile in the point of connection during the fault and the recovery.

7.1 Wind farm modeling

Wind farm models may be built with different detail levels ranging from one-to-one modeling or by an aggregated model that consists of one or few equivalent wind turbines and an equivalent of the internal network. The aggregated model includes: wind turbine units, compensating capacitors, step-up transformers, etc. Fig. 25 compares the detailed and the aggregated models.

The aggregated model can be used to verify a wind installation according to PVVC when all the wind turbines that form the wind installation are of the same type. If a wind installation is formed by different wind turbines, aggregated model can be done grouping the wind turbines of the same type.

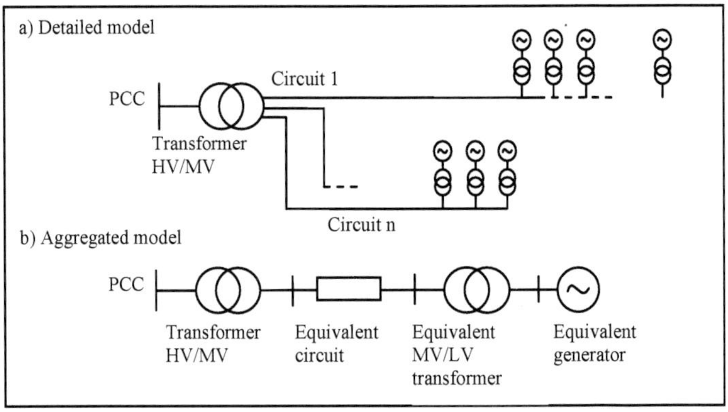

Fig. 25. Wind farm modeling.

Considering identical machines the equivalent generator rating is obtained adding all the machine ratings (García-Gracia et al, 2008):

$$S_{eq} = \sum_{i=1}^{n} S_i \qquad P_{eq} = \sum_{i=1}^{n} P_i \tag{11}$$

where S_i is the *i-th* generator apparent power and P_i is the *i-th* real power.

The inertia H_{eq} and the stiffness coefficient K_{eq} of the equivalent generator are calculated as follows:

$$H_{eq} = \sum_{i=1}^{n} H_i \qquad K_{eq} = \sum_{i=1}^{n} K_i \tag{12}$$

and the size of the equivalent compensating capacitors is given by:

$$C_{eq} = \sum_{i=1}^{n} C_i \tag{13}$$

When the aggregated model is used, the difference between the results obtained by the two models must be negligible. Fig. 26 and Fig. 27 show the results obtained in a example wind farm. Fig. 26 shows a comparison between the real power obtained by the simulation of a detalied and aggregated model. The blue line represents the results of the detailed model, the red line the results of the aggregated model and the green line shows the tolerance (10%). Fig. 27 shows the same comparison for the reactive power. In this case the aggregated model can be used because the differences are negligible during the simulation.

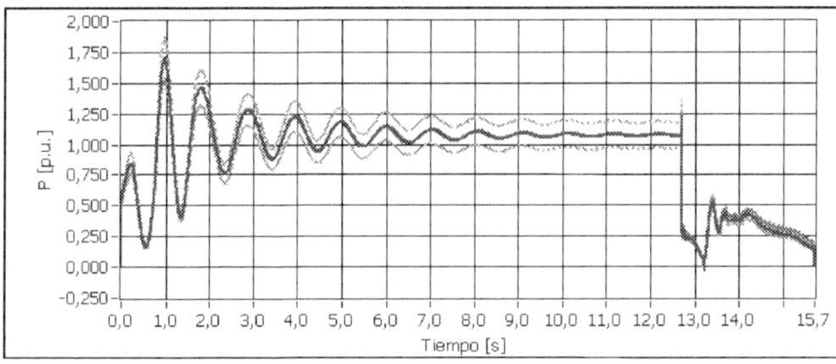

Fig. 26. Real power in the detailed (blue) and the aggregated (red) model.

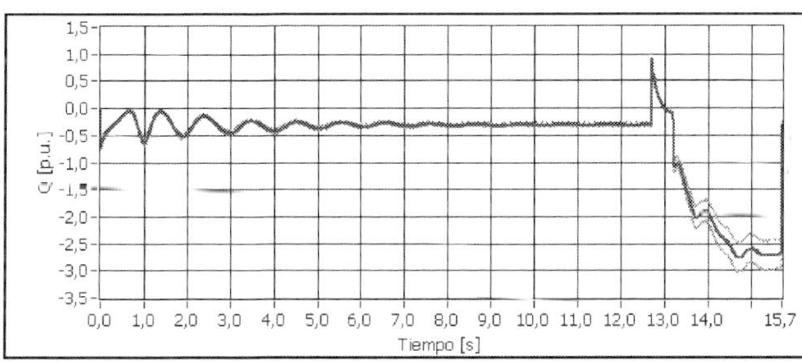

Fig. 27. Reactive power in the detailed (blue) and the aggregated (red) model.

7.2 Modeling wind turbine when there is no available data

Usually, when old installations are going to be verified according to PVVC, there are no available data to model the installation. In these cases, if the rms voltage during the simulation remains above 0.85 p.u , the wind turbines can be represented by a library model that takes into account the generator protections that would disconnect the installation.

If the requirements to use library models are not fulfilled, that is, the voltage falls bellow 0.85 p.u. during the simulation, validated models of the dynamic parts of the wind installation (wind turbines and FACTS) must be provided by the manufacturers. The model validation must be done according PVVC (see section 6).

7.2.1 Characteristics of the wind turbine library

Depending on the wind turbine technology, different models must be used.

For squirrel cage induction generator, a fifth order model must be used. If there are manufacturer data available, the behaviour in rated conditions must be checked with a tolerance of 10% for real and reactive power.

If there are not available data, PVVC establishes the data from Table 11, and the rest of the parameters must be calculated to obtain the rated characteristics of the modelled machine.

Stator resistance (p.u.)	0.005 - 0.007
Rotor resistance (p.u.)	0.005 - 0.007
Stator leakage reactance (p.u.)	0.1 - 0.15
Rotor leakage reactance (p.u.)	0.04 - 0.06
Magnetizing reactance (p.u.)	4 - 5

Table 11. Squirrel cage induction generator characteristic parameters.

If there are no manufacturer data for the wind turbine inertia, the value to model the wind turbine is H = 4 s.
For the doubly fed induction generator, the simplyfied model must take into account the rotor dynamics, to determine the overcurrent tripping of the wind turbine during voltage dips.
Finally, the simplified model of the full converter generator consists of a constant current source.

7.3 Evaluation of the wind installation response

Once the system has been modelled, the evaluation simulations must be performed. The test categories and the operation point prior the voltage dip in the verification process are the same of the in-field test, shown in Table 3 and Table 6 (section 5.2), but, in the simulation, the reactive power before the voltage dip must be zero.
In the simulation results, the next requirements must be checked:

1. Continuity of supply. The wind farm must withstand the dips without disconnection. The simulation model must include the protections that determine the disconnection of the wind turbines. As has been shown in section 7.1, there are two possibilities for the wind farm modeling:
   - Detailed model (without aggregation). In this case, the continuity of supply is guaranteed if the real power of the disconnected wind turbines during the simulation does not exceed the 5% of the real power before the dip.
   - Aggregated model. In this case, the continuity of supply is guaranteed if the equivalent generator remains connected during the simulation of the dips.
2. Voltage and current levels at the WTG terminals. Before verification simulations, a no load simulation must be done, in order to check that the depth and the duration of the simulation of the voltage dips fulfil the PVVC requirements (see section 5.2).
   During the simulation of the four categories shown in Table 3, voltage and current values in each phase must be measured and recorded with a sampling frequency at least of 5 kHz.
   If a library model is used the voltage must remain above 0.85 p.u. during the simulation
3. Real and reactive power exchanges as described in OP 12.3. The power exchanges must fulfil the requirements shown in Table 12 and Table 13.

The definition of the different zones is shown in Fig. 17.

Three phase faults	OP 12.3 requirements
ZONE A	
Net consumption Q < 60% Pn (20 ms)	-0.6 p.u.
ZONE B	
Net consumption P < 10% Pn (20 ms)	-0.1 p.u.
Average I_r/I_{tot}	0.9 p.u.
ZONE C	
Net consumption E_r < 60% Pn * 150 ms	-90 ms*p.u.
Net consumption I_r < 1.5 I_n (20 ms)	-1.5 p.u.

Table 12. Power and energy requirements for three phase voltage dips in the General Verification Process.

Two phase faults	OP 12.3 requirements
ZONE B	
Net consumption E_r < 40% Pn * 100 ms	-40 ms*p.u.
Net consumption Q < 40% Pn (20 ms)	-0.4 p.u.
Net consumption E_a < 45% Pn * 100 ms	-45 ms*p.u.
Net consumption P < 30% Pn (20 ms)	-0.3 p.u.

Table 13. Power and energy requirements for isolated two phase voltage dips in the General Verification Process.

8. References

Amarís, H. (2007). Power Quality Solutions for Voltage dip compensation at Wind Farms, *Power Engineering Society General Meeting*, 2007. IEEE , Issue Date: 24-28 June 2007

Asociación Empresarial Eólica (AEE). Procedure for verification validation and certification of the requirements of the PO 12.3 on the response of wind farms in the event of voltage dips. November 2007. http://www.aeeolica.es/doc/privado/pvvc_v3_english.pdf

Bundesministerium der Ordinance on system services by wind energy plants (system services ordinance - SDLWindV), 03 July 2009, published in the Federal Law Gazette 2009, Part I, No. 39

REE. (2006). Requisitos de respuesta frente a huecos de tensión de las instalaciones de producción de Régimen Especial, Procedimiento de Operación 12.3. Red Eléctrica de España. October 2006.

Fördergesellschaft Windenergie und andere Erneuerbare Energien (FGW), Technical Guidelines for Power Generating Units. Part 3. Determination of electrical characteristics of power generating units to MV, HV and EHV grids, Revision 20, 01.10.2009

Fördergesellschaft Windenergie und andere Erneuerbare Energien (FGW), Technical Guidelines for Power Generating Units. Part 4. Requirements for modelling and validation of simulation models of the electrical characteristics of power generating units and systems, Revision 4, 01.10.2009

Fördergesellschaft Windenergie und andere Erneuerbare Energien (FGW), Technical Guidelines for Power Generating Units. Part 8. Certification of the electrical characteristics of power generating units and systems in the medium., high- and highest-voltage grids, Revision 1, 01.10.2009

Hingorani, N. G. & Gyugyi, L. (1999). *Understanding FACTS: concepts and technology of flexible AC transmission system*. Wiley-IEEE Press, 1999

Gamesa Eólica, S.A. Patent WO/2006/108890. Voltage sag generator device. Sag-swell and outage generator for performance test of custom power devices

Gamesa Innovation and Technology, S.L. Patent WO/2006/106163. Low-Voltage dip generator device.

García-Gracia, M.; Comech, M.P.; Sallán, J. & Llombart, A. (2008) Modelling wind farms for grid disturbance studies. *Renew Energy* (2008), doi:10.1016/j.renene.2007.12.007.

García-Gracia. M.; Comech, M.P.; Sallán. J.; Lopez-Andía, D. & Alonso, O. (2009). Voltage dip generator for wind energy systems up to 5 MW, *Applied Energy*, 86 (2009) 565–574, doi:10.1016/j.apenergy.2008.07.006

Jauch, C.; Sørensen, P.; Norhem, I. & Rasmussen, C. (2007). Simulation of the impact of wind power on the transient fault behaviour of the Nordic power system. *Electric Power Syst Res* 2007;77:135-44.

Khadkikar, V. ; Aganval, P.; Chandra, A.; Bany A.O. & Nguyen T.D. (2004). A Simple New Control Technique For Unified Power Quality Conditioner (UPQC), *11th International Conference on Harmonics and Quality of Power*

López, J.; Gubía, E.; Olea, E.; Ruiz, J. & Luis Marroyo, L. (2009). Ride Through of Wind Turbines With Doubly Fed Induction Generator Under Symmetrical Voltage Dips. *IEEE Transactions On Industrial Electronics*, Vol. 56, No. 10, Oct 2009

Molinas, M.; Suul, J.A. & Undeland, T. (2008). Low Voltage Ride Through of Wind Farms With Cage Generators: STATCOM Versus SVC. *IEEE Transactions On Power Electronics*, Vol. 23, No. 3, May 2008

Morren, J. & de Haan, S.W.H (2005) .Ridethrough of wind turbines with doubly fed induction generators during a voltage dip. *IEEE Trans. Energy Convers.* vol. 20, no. 2, pp. 435-441, Jun. 2005

Morren, J. & de Haan, S.W.H. (2007) Short-Circuit current of wind turbines with doubly fed induction generator. *IEEE Trans. On Energy convers*, vol. 22, no. 1, march 2007

Muyeen, S.M.; Takahashi, R.; Murata, T.; Tamura, J.; Ali, M.H.; Matsumura, Y.; Kuwayama, A. & Matsumoto, T. (2009). Low voltage ride through capability enhancement of wind turbine generator system during network disturbance. *IET Renew. Power Gener.*, 2009, Vol. 3, No. 1, pp. 65–74, ISSN 1752-1416

Muyeen, S.M. & Rion Takahashi, R. (2010). A Variable Speed Wind Turbine Control Strategy to Meet Wind Farm Grid Code Requirements. *IEEE Transactions On Power Systems*, Vol. 25, No. 1, Feb 2010 331-340

Niiranen J. Experiences on voltage dip ride through factory testing of synchronous and doubly fed generator drives. *11th European Conference on Power Electronics and Applications*. Dresden 2005

Rodríguez, J.M.; Fernández, J.L.; Beato, D.; Iturbe, R.; Usaola, J.; Ledesma, P. (2002). Incidence on power system dynamics of high penetration of fixed speed and doubly fed wind energy systems: study of the Spanish case. *IEEE Trans Power Syst 2002*;17(4):1089-95

Wizmar Wahab, S.; and Mohd Yusof. A. (Elektrika Voltage Sag and Mitigation Using Dynamic Voltage Restorer (DVR) System. VOL. 8, NO. 2, 2006, 32-37

3

Active and Reactive Power Formulations for Grid Code Requirements Verification

Vicente León-Martínez and Joaquín Montañana-Romeu
Universidad Politécnica de Valencia
Spain

1. Introduction

Wind power penetration has reached important levels in several European, American and other world countries. Wind electric energy production in some countries is comparable with that obtained through the nuclear and other conventional energies, thus System Operators in many nations have established wind farms grid codes in order to remain grid stability. Grid code requirements have been developed in response to the technical and regulatory necessities in each country; so there are a great variety of wind farms connection requirements. However, all grid codes have in common some quantities such as voltage, frequency and active and reactive powers and currents must be verified.

In other hand, grid code requirements do not specify which active and reactive power and current formulations must be used. A lot of power approaches can be used. Several recently established approaches consider active and reactive phenomena must be analyzed by the fundamental-frequency, positive-sequence voltages and currents; this is because these last quantities determinate generators working and electromechanical stability. The IEEE Standard 1459-2010 explicitly holds one of these theories, due to A.E. Emanuel. The p-q-r theory, developed by Akagi and others, also establishes fundamental-frequency, positive-sequence active and reactive powers. The Unified Theory described in this Chapter gives one more step in front of the two above mentioned theories and decomposes fundamental-frequency, positive-sequence active and reactive powers and currents into two quantities: a) due to the active and reactive loads and b) caused by the unbalances. According to the Unified Theory unbalances can originate additional active and reactive powers and currents which can have the same or different sign of those due to active and reactive loads and, therefore, total active and reactive powers and currents can be increased or decreased. This active and reactive powers and currents decomposition can deliver important complementary information for verifying accomplishment of the grid code requirements and to regulate wind generators in order to win without disconnection transitory perturbations, such as voltage dips.

In this Chapter, the two above indicated fundamental-frequency, positive-sequence active and reactive components of powers and currents are expressed and their properties are established. Formulations of these quantities are applied on actual wind farms to verify some European Grid Code requirements, focusing on the Spanish grid code, and their results are compared with those obtained from other power approaches.

Conclusions show that power and current formulations established in this Chapter are important tools to analyze wind farms working in normal operation and in presence of transitory disturbances, and these formulations can be proposed for a future grid code harmonisation.

2. Active and reactive powers and currents formulations applied to wind farms

Figure 1 schematically shows the equivalent circuit of a wind generator connected to the grid (represented by a delta-connected load). Phases of the wind generator are star-connected and there is no neutral wire. Active and reactive phenomena in these power systems do not depend on the zero-sequence voltages and, thus, any artificial ground can be chosen to measure phase voltages at the point of common coupling (PCC).

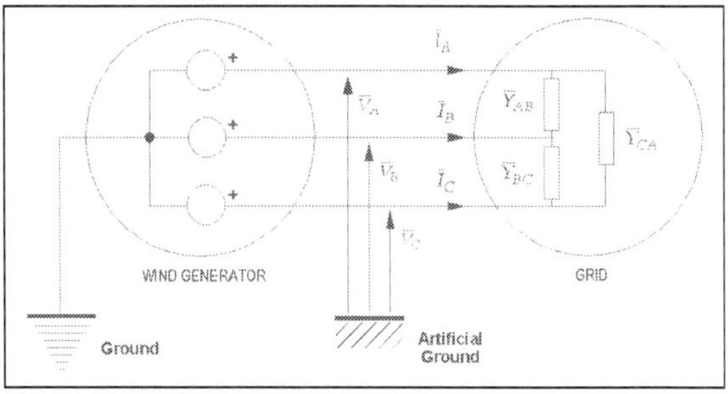

Fig. 1. Equivalent circuit of a wind generator connected to the grid

Active and reactive phenomena in that power system are analyzed and their characteristic quantities are formulated in this section using the Unified Theory (León et al., 2001). Traditional active and reactive powers included in the IEEE Standard 1459-2010 will be expressed at last of this section in order to compare the results obtained with these mentioned approaches applied on data registered in actual wind farms, in other sections.

2.1 Active and reactive phenomena according to the unified theory

Unified Theory (León et al., 2001) establishes the active and the reactive phenomena occur because the fundamental positive-sequence voltages and currents. This consideration also is implicitly established by the p-q-r theory (Kim et al., 2002) and Emanuel's theory, included in the IEEE Standard 1459-2010. Importance of the fundamental-frequency positive-sequence quantities is they determinate the main magnetic field and the useful torque of the wind generators and, consequently, the adequate working and stability of those machines. Contribution of the Unified Theory with respect to the two above mentioned approaches is active and reactive currents and powers have been decomposed into two components: (a) due to the loads and (b) caused by the unbalances (León et al., 2007; 2009). These new quantities established by the Unified Theory give better and greater information about the manifesting phenomena, which can be applied to analyze wind generators working.

2.1.1 Unified theory's active and reactive currents

Let's consider the equivalent circuit of a wind-generator connected to the grid, represented in fig.1. Fundamental-frequency voltages obtained at the point of common coupling (PCC)

by Fourier's analysis are unbalanced, in general, and their CRMS line to line values ($\overline{V}_{AB}, \overline{V}_{BC}, \overline{V}_{CA}$) can be decomposed into the positive-sequence ($\overline{V}_{AB+}$) and the negative-sequence ($\overline{V}_{AB-}$) components, by Stokvis-Fortescue:

$$\begin{aligned} \overline{V}_{AB} &= \overline{V}_{AB+} + \overline{V}_{AB-} \\ \overline{V}_{BC} &= \overline{V}_{BC+} + \overline{V}_{BC-} = a^2\,\overline{V}_{AB+} + a\,\overline{V}_{AB-} \\ \overline{V}_{CA} &= \overline{V}_{CA+} + \overline{V}_{CA-} = a\,\overline{V}_{AB+} + a^2\,\overline{V}_{AB-} \end{aligned} \tag{1}$$

expressions where $a = 1/120^o$ and the voltage symmetrical components are obtained as:

$$\begin{aligned} \overline{V}_{AB+} &= \tfrac{1}{3}(\overline{V}_{AB} + a\overline{V}_{BC} + a^2\,\overline{V}_{CA}) = V_{AB+}\,|\underline{\alpha_+} \\ \overline{V}_{AB-} &= \tfrac{1}{3}(\overline{V}_{AB} + a^2\,\overline{V}_{BC} + a\overline{V}_{CA}) = V_{AB-}\,|\underline{\alpha_-} \end{aligned} \tag{2}$$

Load phase currents be expressed in function of those voltage symmetrical components and the load admittances ($\overline{Y}_{AB}, \overline{Y}_{BC}, \overline{Y}_{CA}$):

$$\begin{aligned} \overline{I}_{AB} &= \overline{Y}_{AB} \cdot \overline{V}_{AB} = \overline{Y}_{AB} \cdot (\overline{V}_{AB+} + \overline{V}_{AB-}) \\ \overline{I}_{BC} &= \overline{Y}_{BC} \cdot \overline{V}_{BC} = \overline{Y}_{BC} \cdot (a^2\,\overline{V}_{AB+} + a\,\overline{V}_{AB-}) \\ \overline{I}_{CA} &= \overline{Y}_{CA} \cdot \overline{V}_{CA} = \overline{Y}_{CA} \cdot (a\,\overline{V}_{AB+} + a^2\,\overline{V}_{AB-}) \end{aligned} \tag{3}$$

These currents are unbalanced, in general, and thus their symmetrical components are, by Stokvis-Fortescue:

$$\begin{aligned} \overline{I}_{AB+} &= \overline{Y}_+ \cdot \overline{V}_{AB+} + \overline{Y}_i \cdot \overline{V}_{AB-} \\ \overline{I}_{AB-} &= \overline{Y}_h \cdot \overline{V}_{AB+} + \overline{Y}_+ \cdot \overline{V}_{AB-} \\ \overline{I}_{ABo} &= \overline{Y}_i \cdot \overline{V}_{AB+} + Y_h \cdot \overline{V}_{AB-} \end{aligned} \tag{4}$$

where subscripts (+), (-) and (o), respectively denote positive-, negative- and zero-sequence components, and the admittances are:

- Positive admittance,

$$\overline{Y}_e = \tfrac{1}{3}(\overline{Y}_{AB} + \overline{Y}_{BC} + \overline{Y}_{CA}) = Y_e\,|\underline{-\alpha_e} \tag{5}$$

- Basic unbalance admittance for the negative-sequence,

$$\overline{Y}_i = \tfrac{1}{3}(\overline{Y}_{AB} + a^2\,\overline{Y}_{BC} + a\overline{Y}_{CA}) = Y_i\,|\underline{-\alpha_i} \tag{6}$$

- Basic unbalance admittance for the positive-sequence,

$$\overline{Y}_h = \tfrac{1}{3}(\overline{Y}_{AB} + a\overline{Y}_{BC} + a^2\,\overline{Y}_{CA}) = Y_h\,|\underline{-\alpha_h} \tag{7}$$

Positive admittance ($\overline{Y}_e$) is the admittance of the equivalent balanced load which absorbs the same active and reactive powers that the real unbalanced load when are supplied with the fundamental-frequency positive sequence voltages. Basic unbalance admittance for the

negative-sequence ($\overline{Y}_i$) denotes the increasing of the fundamental positive-sequence currents due to the negative-sequence voltage effects. Basic unbalance admittance for the positive-sequence ($\overline{Y}_h$) defines the increasing of the fundamental negative-sequence currents due to the positive-sequence voltage effects.
Line to artificial-ground voltages ($\overline{V}_A, \overline{V}_B, \overline{V}_C$) at the PCC of the circuit showed in fig. 1 have the following fundamental positive- and negative-sequence components, by Stokvis-Fortescue:

$$\overline{V}_{A+} = \frac{\overline{V}_{AB+}}{\sqrt{3}}\,|{-30^\circ} \qquad \overline{V}_{A-} = \frac{\overline{V}_{AB-}}{\sqrt{3}}\,|30^\circ \tag{8}$$

Fundamental positive-sequence line currents ($\overline{I}_A, \overline{I}_B, \overline{I}_C$) supplied by the wind-generator showed in fig. 1 are unbalanced have the following general expression, from (4) and (8):

$$\overline{I}_{A+} = \sqrt{3}\,\overline{I}_{AB+}\,|{-30^\circ} = 3\overline{V}_{A+}\cdot(\overline{Y}_e + \overline{\delta}_u \cdot \overline{Y}_i) \tag{9}$$

where

$$\overline{\delta}_u = \frac{\overline{V}_{AB-}}{\overline{V}_{AB+}} = \delta_u\,|\alpha_- - \alpha_+ \tag{10}$$

is the unbalance degree of the phase to phase voltages at the PCC.
From (9), two components of the fundamental positive-sequence line currents may be established: active and reactive. Active fundamental positive-sequence line current ($\overline{I}_{Aa+}$) has the following general expression:

$$\begin{aligned}\overline{I}_{Aa+} &= 3\overline{V}_{A+}\cdot(Y_e \cdot \cos\alpha_e + \delta_u \cdot Y_i \cdot \cos(\alpha_- - \alpha_+ - \alpha_i)) = \\ &= 3\overline{V}_{A+}\cdot(G_e + \delta_u \cdot Y_i \cdot \cos(\alpha_- - \alpha_+ - \alpha_i))\end{aligned} \tag{11}$$

being $G_e = Y_e \cdot \cos\alpha_e$ the load positive conductance, the real part of the positive admittance ($\overline{Y}_e$). The above current is 0° dephased with the fundamental positive-sequence phase to ground voltage ($\overline{V}_{A+}$) and it transfers the useful power (positive-sequence active power, P_+) produced by the wind-generator. Active fundamental positive-sequence line current may be decomposed into two components too, as it is appreciated from (11):

$$\begin{aligned}\overline{I}_{Aaa+} &= 3\;Y_e \cdot \cos\alpha_e \cdot \overline{V}_{A+} = 3\,G_e\,\overline{V}_A \\ \overline{I}_{Aau+} &= 3\,\delta_u \cdot Y_i \cdot \cos(\alpha_- - \alpha_+ - \alpha_i)\cdot\overline{V}_{A+}\end{aligned} \tag{12}$$

First component of the active fundamental positive-sequence line currents, $\overline{I}_{Aaa+}$, transfers the active power in the best efficiency and power quality conditions (P_{a+}), i.e., when voltages are sinusoidal and balanced, with positive-sequence. Second component, $\overline{I}_{Aau+}$, characterizes the increasing (positive or negative) of positive-sequence active power caused by the voltage and load (grid) unbalances (P_{u+}).
Reactive fundamental positive-sequence line current ($\overline{I}_{Ar+}$) is the component of $\overline{I}_{A+}$ 90° dephased with respect to $\overline{V}_{A+}$, which transfers the positive-sequence reactive power (Q_+). General expression of this current is, from (9):

$$\begin{aligned}\overline{I}_{Ar+} &= j\,3\overline{V}_{A+}\cdot(-Y_e\cdot\sin\alpha_e+\delta_u\cdot Y_i\cdot\sin(\alpha_- -\alpha_+ -\alpha_i)) = \\ &= j\,3\overline{V}_{A+}\cdot(\mp B_e+\delta_u\cdot Y_i\cdot\sin(\alpha_- -\alpha_+ -\alpha_i))\end{aligned} \tag{13}$$

where $B_e = Y_e\cdot\sin\alpha_e$ is the load positive susceptance, the imaginary part of the positive admittance ($\overline{Y}_e$) Reactive fundamental positive-sequence line current also holds two components:

$$\begin{aligned}\overline{I}_{Arr+} &= -j\,3\,Y_e\cdot\sin\alpha_e\cdot\overline{V}_{A+} = \mp j\,3\,B_e\,\overline{V}_{A+} \\ \overline{I}_{Aru+} &= j\,3\,\delta_u\cdot Y_i\cdot\sin(\alpha_- -\alpha_+ -\alpha_i)\cdot\overline{V}_{A+}\end{aligned} \tag{14}$$

First component, $\overline{I}_{Arr+}$, transfers the positive-sequence reactive power with balanced voltages (Q_{r+}); thus, this current delivers the load reactive power (negative sign of this quantity in (14) corresponds with inductive loads and positive sign is for capacitive loads). Second component, $\overline{I}_{Aru+}$, represents the increasing (positive or negative) of the reactive power caused by the voltage and load (grid) unbalances (Q_{u+}).

2.1.2 Unified theory's active and reactive powers

Fundamental positive-sequence complex power supplied by the wind generator showed in fig. 1 is expressed as:

$$\overline{S}_+ = 3\,\overline{V}_{A+}\cdot\overline{I}^*_{A+} = 9V^2_{A+}\cdot(\overline{Y}^*_e+\overline{\delta}^*_u\cdot\overline{Y}^*_i) = P_+ + \overline{Q}_+ \tag{15}$$

Positive-sequence active power (P_+) is the real part of the above quantity and it characterizes the direct torque applied to the axis of the wind generator. This quantity has two components, due to the active loads (P_{a+}) and caused by the unbalances (P_{u+}):

$$\begin{aligned}P_+ &= 3\,\overline{V}_{A+}\cdot\overline{I}^*_{Aa+} = 9V^2_{A+}\cdot(G_e+\delta_u\cdot Y_i\cdot\cos(\alpha_+ -\alpha_- +\alpha_i)) = P_{a+}+P_{u+} \\ P_{a+} &= 3\,\overline{V}_{A+}\cdot\overline{I}^*_{Aaa+} = 9\,G_e\,V^2_{A+} \\ P_{u+} &= 3\,\overline{V}_{A+}\cdot\overline{I}^*_{Aau+} = 9\,\delta_u\cdot Y_i\cdot\cos(\alpha_+ -\alpha_- +\alpha_i)\cdot V^2_{A+}\end{aligned} \tag{16}$$

P_{a+} is the positive-sequence active power supplied by the wind-generator under positive-sequence balanced voltages; thus, it may be defined as the positive-sequence active power due to the load consumptions. This quantity measures the active power which is transformed under the best efficiency and power quality conditions. P_{u+} represents the increasing of the positive-sequence active power produced by the voltage and load unbalances. Last quantity identifies the poor power quality in the power system, since it occurs when there are voltage unbalances, and it may have the same or different sign that P_{a+}, so it increases or decreases the total positive-sequence active power (P_+).

Positive-sequence reactive power (Q_+) is the module of the imaginary part of the positive-sequence complex power. Expressed in complex notation, this quantity has the following formulation:

$$\begin{aligned}\overline{Q}_+ &= 3\,\overline{V}_{A+}\cdot\overline{I}^*_{Ar+} = j\,9V^2_{A+}\cdot(Y_e\cdot\sin\alpha_e+\delta_u\cdot Y_i\cdot\sin(\alpha_+ -\alpha_- +\alpha_i)) = \\ &- j\,9V^2_{A+}\cdot(\pm B_e+\delta_u\cdot Y_i\cdot\sin(\alpha_+ -\alpha_- +\alpha_i)) = \overline{Q}_{r+}+\overline{Q}_{u+}\end{aligned} \tag{17}$$

Positive-sequence reactive power characterizes the main magnetic field of the wind-generator and it holds two components, due to the reactive loads ($\bar{Q}_{r+}$) and caused by the unbalances ($\bar{Q}_{u+}$):

$$\begin{aligned} \bar{Q}_{r+} &= 3\,\bar{V}_{A+}\cdot\bar{I}^{*}_{Arr+} = j\,9\;Y_e\cdot\sin\alpha_e\cdot V_{A+}^2 = \pm j\,9\,B_e\,V_{A+}^2 \\ \bar{Q}_{u+} &= 3\,\bar{V}_{A+}\cdot\bar{I}^{*}_{Aru+} = j\,9\,\delta_u\cdot Y_i\cdot\sin(\alpha_+ - \alpha_- + \alpha_i)\cdot V_{A+}^2 \end{aligned} \tag{18}$$

Q_{r+} is the positive-sequence reactive power supplied by the wind-generator under positive-sequence balanced voltages. This quantity determinates the reactive power established under the best efficiency and power quality conditions. Q_{u+} defines the increasing of the positive-sequence active power produced by the voltage and load unbalances. This quantity identifies the poor power quality in the power system, since it occurs when there are voltage unbalances, and it may have the same or different character (inductive or capacitive) that Q_{r+}, and thus it can increase or decrease the positive-sequence reactive power , Q_+ .

2.2 Active and reactive phenomena according to the Spanish Grid Code

Active and reactive currents and powers are not explicitly formulated in the Spanish Grid Code (O.P. 12.3); however, traditional formulations of these quantities can be implicitly appreciated in the grid code text, such as will be seen in the next section. Those active and reactive formulations are obtained from Budeanu´s approach, applied to sinusoidal circuits, and they are included into the IEEE Standard 1459-2010.

Active and reactive currents supplied by the wind-generator ($\bar{I}_{az}, \bar{I}_{rz}$, z=A,B,C) are the traditionally known fundamental-frequency line current 0° and ± 90° respectively dephased with respect to its fundamental phase voltage ($\bar{V}_z$),

$$\bar{I}_{az} = G_z\cdot\bar{V}_z = \frac{P_z}{V_z^2}\,\bar{V}_z \qquad \bar{I}_{rz} = \bar{B}_z\cdot\bar{V}_z = \mp j\,\frac{Q_z}{V_z^2}\,\bar{V}_z \tag{19}$$

Active current transfers the active power of each phase (P_z) and reactive current delivers the reactive power of the correspondent phase (Q_z).

Active and reactive powers supplied by the wind-generator, according to the Spanish Grid Code implicitly proposes, are the well-known active and reactive powers for sinusoidal three-phase circuits:

$$\begin{aligned} P &= \sum_{z=A,B,C} P_z = \bar{V}_A\cdot\bar{I}^{*}_{aA} + \bar{V}_B\cdot\bar{I}^{*}_{aB} + \bar{V}_C\cdot\bar{I}^{*}_{aC} \\ \bar{Q} &= \sum_{z=A,B,C} \bar{Q}_z = \bar{V}_A\cdot\bar{I}^{*}_{rA} + \bar{V}_B\cdot\bar{I}^{*}_{rB} + \bar{V}_C\cdot\bar{I}^{*}_{rC} \end{aligned} \tag{20}$$

Positive-sequence active and reactive powers (P_+, Q_+) described in the before section are respectively included in the above quantities, but also active and reactive powers expressed by (20) contain quantities due to the fundamental-frequency negative-sequence voltages and currents (P_-, Q_-).

3. Grid code requirements

Grid codes established by the different countries provides the minimum operation and security requirements of the wind farms installations connected to the Electric Network in order to guarantee the supply continuity in presence of voltage dips. The Spanish Operation Procedure O.P. 12.3, which constitutes the present Spanish Grid Code, establishes wind farms and all their components must be able to withstand, without disconnection, transient voltage dips at the grid point of common coupling caused by three-phase, two-phase and single-phase faults within the area described by the voltage-time characteristic showed in fig.2a. That characteristic or LVRT (Low Voltage Ride Through) requirements has been recently modified by the draft of the Spanish Operation Procedure O.P. 12.2 by increasing the allowed depth of the voltage drop up to zero during the first 150 ms after the beginning of the disturbance (fig.2b), similar to the LVRT requirements of the German Grid Code from E.ON Netz, represented in fig.2c.

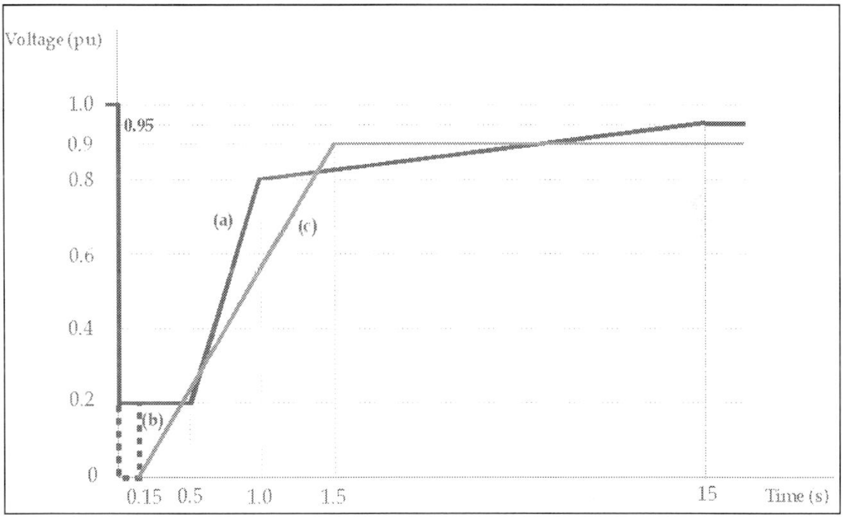

Fig. 2. Low Voltage Ride Through requirements: (a) Spanish O.P. 12.3, (b) Spanish O.P. 12.2 (draft), (c) E.ON Netz

3.1 Reactive power requirements

The present Spanish Grid Code (O.P. 12.3) prescribes that reactive power consumptions are not allowed in the wind farm installations at the point of common coupling with the grid during the voltage dip and the following clearance fault and voltage recovery. However, some reactive power consumptions lower than 60% of the registered rated power in each cycle (20 ms) may be allowed during just the 150 ms after the beginning of three-phase balanced voltage dips and the 150 ms after its clearance (fig.3a). These admitted periods of reactive power consumptions will be reduced in the future Spanish Grid Code (O.P. 12.2) to 40 ms after the beginning of the fault and 80 ms after the voltage recovery and clearance fault (fig.4a).

For unbalanced single-phase and two-phase voltage dips (fig.3b), some unspecified reactive power consumptions are allowed during the 150 ms after the beginning of the fault (80 ms according to the O.P. 12.2, fig.4b) and the 150 ms after the voltage recovery (80 ms according to the O.P. 12.2, fig.4b). But, some reactive power consumptions lower than 40% of the registered rated powers are admitted during all disturbance duration for periods lower than 100 ms.
Reactive power for unbalanced faults is defined by the present Spanish Grid Code like the sum of the reactive powers supplied to each grid phases, i.e., such as it is expressed by (20).
E.ON German Grid Code establishes grid voltages must be supported during the transient voltage dips by supplying the necessary reactive power, with a limit of the wind farm registered rated power.

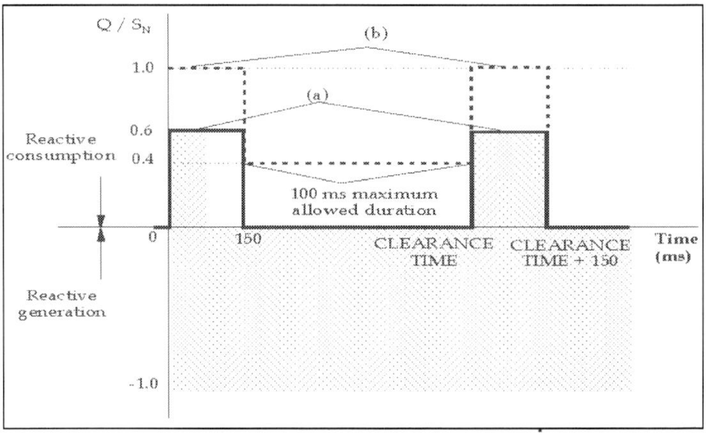

Fig. 3. Reactive power requirements according to the O.P. 12.3: (a) Balanced voltage dips; (b) unbalanced voltage dips

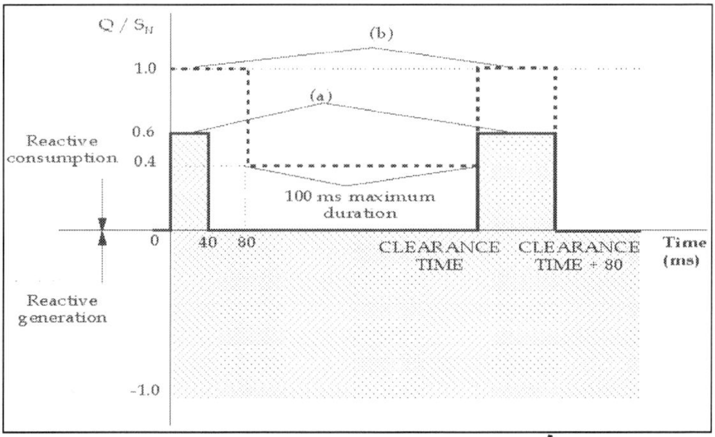

Fig. 4. Reactive power requirements according to the O.P. 12.2: (a) Balanced voltage dips; (b) unbalanced voltage dips

3.2 Active power requirements

The O.P. 12.3 and the draft of the O.P. 12.2 establish no active power consumptions are allowed during the fault and the voltage recovery period. However, some momentary active power consumptions are allowed by both Operation Procedures during the fault and the clearance period, such as figs. 5 and 6 respectively show.

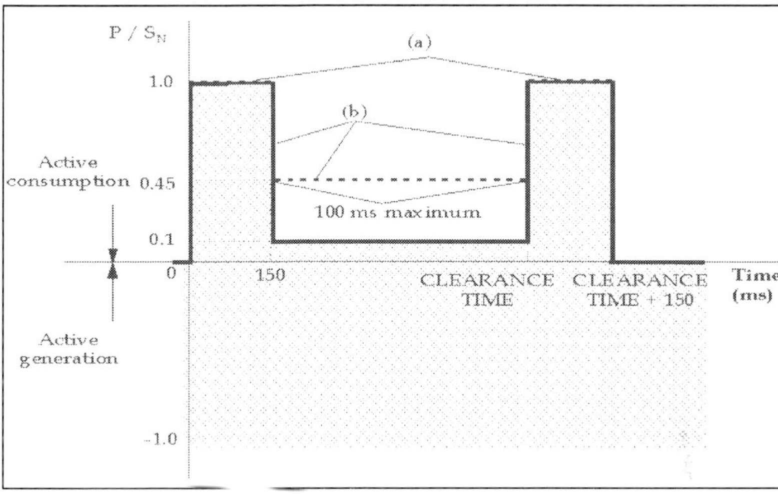

Fig. 5. Active power requirements according to the O.P. 12.3: (a) Balanced voltage dips; (b) unbalanced voltage dips

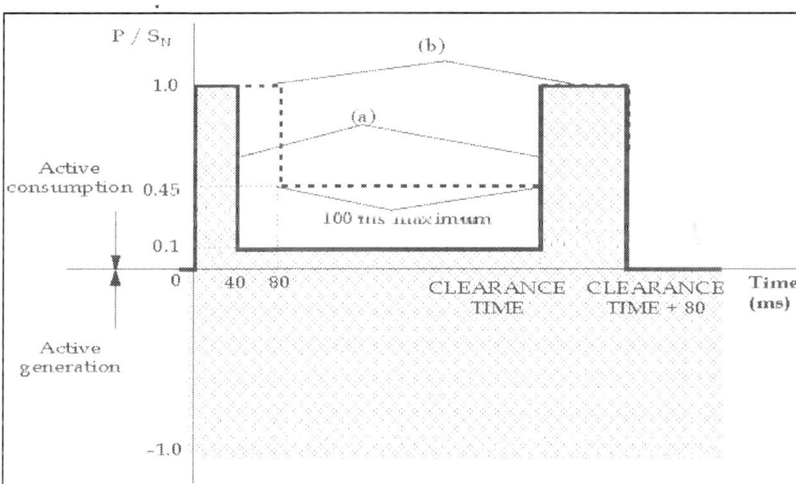

Fig. 6. Active power requirements according to the O.P. 12.2: (a) Balanced voltage dips; (b) unbalanced voltage dips

Active power consumptions lower than 10% of installation registered rated power are admitted during the maintenance of the fault in presence of three-phase balanced voltage dips, while this maximum allowed magnitude is increased up to 45% of registered rated power for unbalanced voltage dips, but only during 100 ms (30% each 20 ms cycle). These active power consumptions referred by the O.P. 12.3 are implicitly defined by (20). The O.P. 12.2 does not express which active power formulation must be used.
German Grid Code is not as exhaustive as the Spanish Grid Code and it specifies wind farms have the ability of active power curtailment with a ramp rate 10% of grid connection per minute.

3.3 Current requirements

Spanish and German Grid Codes require the installation supplies the maximum possible current during the fault maintenance and the voltage recovery period. This current delivery must verify that reactive current is above the minimum unitary values delimited by the lines in fig.7, for each grid code.

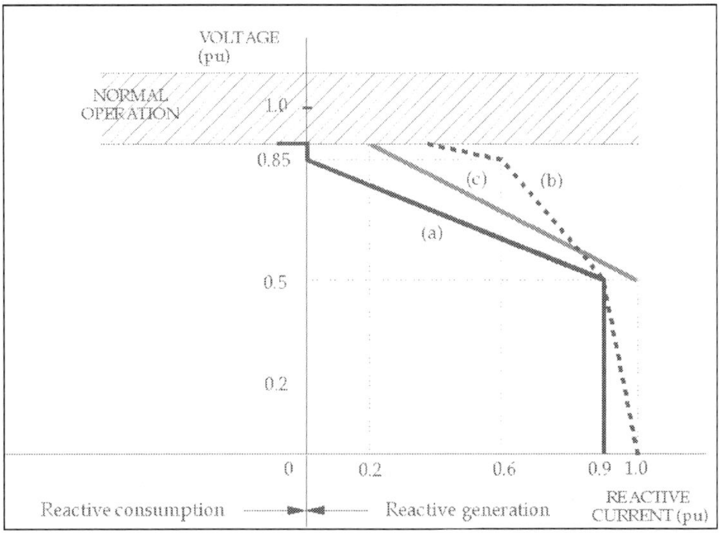

Fig. 7. Minimum admissible values of the reactive current: (a) O.P. 12.3; (b) O.P. 12.2; (c) E.ON Netz

Active current limits (in per unit values) according to the O.P. 12.3 are mathematically expressed in function of the unitary voltage values (V) as:

$$I_a \geq \sqrt{1-\left(1+2{,}57\cdot\left(V-0{,}85\right)\right)^2}$$
$$I_a \leq \sqrt{1-6{,}6\cdot\left(0{,}85-V\right)^2} \qquad \left(0{,}5 \leq V \leq 0{,}85\right) \qquad (21)$$

$$0 \leq I_a \leq 0{,}4359 \qquad \left(0 \leq V \leq 0{,}5\right)$$

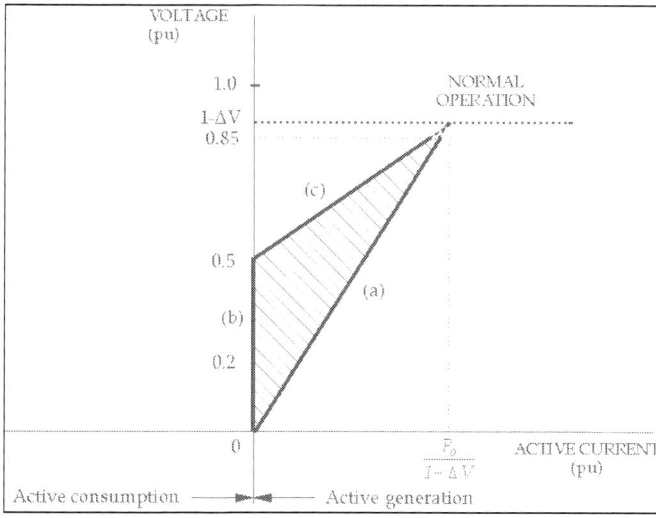

Fig. 8. Active current limits in unitary values during the voltage dip

Active current values according to the O.P. 12.2 must be within the area showed in fig.8. Limits of the active current described in fig.8 are mathematically expressed in unitary values as:

$$
\begin{aligned}
&(a)\quad I_a \leq \frac{P_0}{(1-\Delta V)^2} V && (V \leq 1-\Delta V) \\
&(b)\quad I_a \geq 0 && (V \leq 0{,}5) \\
&(c)\quad I_a \geq \frac{P_0}{(1-\Delta V)(0{,}5-\Delta V)}(V-0{,}5) && (0{,}5 \leq V \leq 1-\Delta V)
\end{aligned}
\qquad (22)
$$

where P_0 is the unitary active power supplied by the installation prior to the disturbance.

4. Practical experiences

Two remarkable events occurred in a Spanish wind farm is used in this section to analyze utility of the active and reactive formulations established in section 2 and their application for verifying grid code requirements. Those events are a three-phase balanced voltage dip and a two-phase voltage dip manifested at the connection point of a 660 kW rated power wind generator, with 690 V phase to phase nominal voltages.

Spanish grid code requirements in their two versions, O.P. 12.2 and O.P. 12.3, were not verified in the three-phase balanced voltage dip (fig. 9) and the installation was finally disconnected, mainly due to an excess of the supplied active current (figs. 10 and 11a). Comparison between active currents measured during the three-phase balanced voltage dip according to the two approaches included in section 2 (figs. 10 and 11a) shows traditional active currents used by the grid codes and fundamental positive-sequence active current have the same evolutions. And the same can be told for the traditional and positive-sequence reactive currents (fig. 12 and 13a). Active and reactive powers show the same tendencies and similar values with both theories (figs 14 and 15, respectively). However, while traditional active and reactive currents have different values in each phase, this one does not occur with the positive-sequence active and reactive currents; thus, the verification process of the grid code requirements is easier using the Unified Theory.

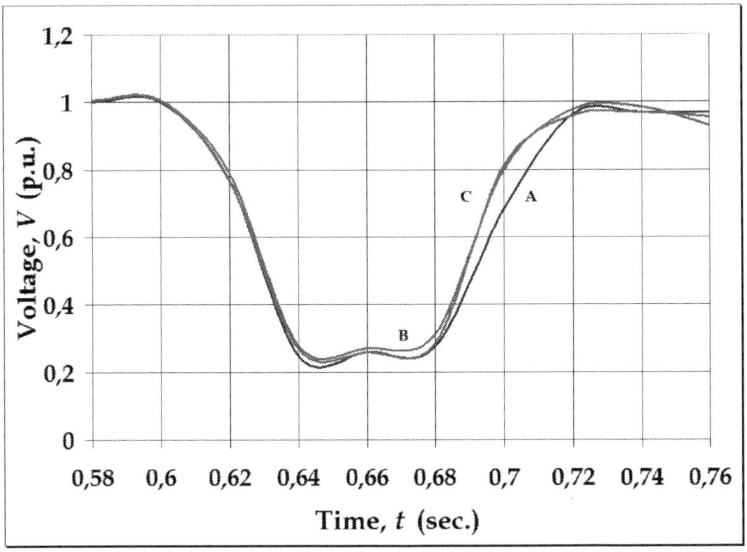

Fig. 9. Three-phase balanced voltage dip

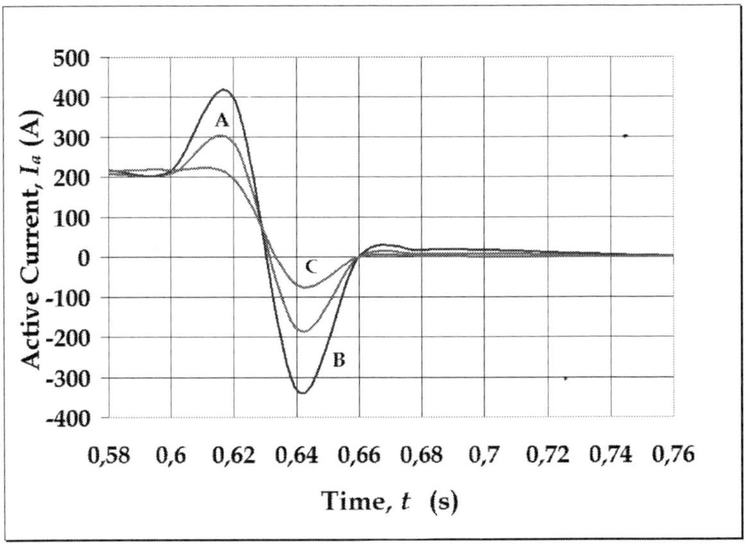

Fig. 10. Phase active currents

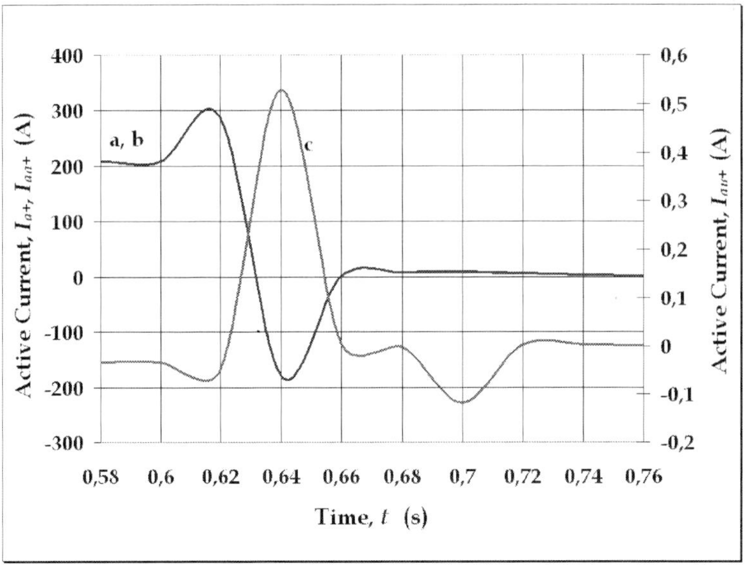

Fig. 11. Unified Theory's active currents: (a) total, (b) due to the active loads, (c) caused by the unbalances

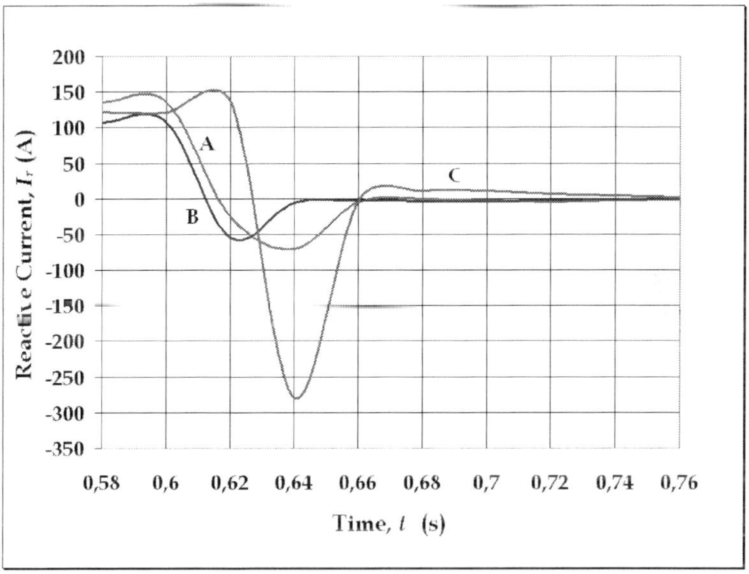

Fig. 12. Phase reactive currents

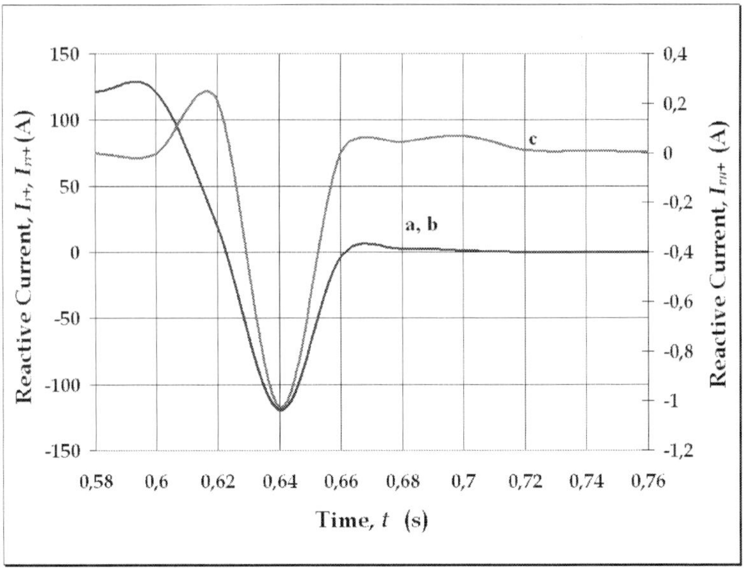

Fig. 13. Unified Theory's reactive currents: (a) total, (b) due to the reactive loads, (c) caused by the unbalances

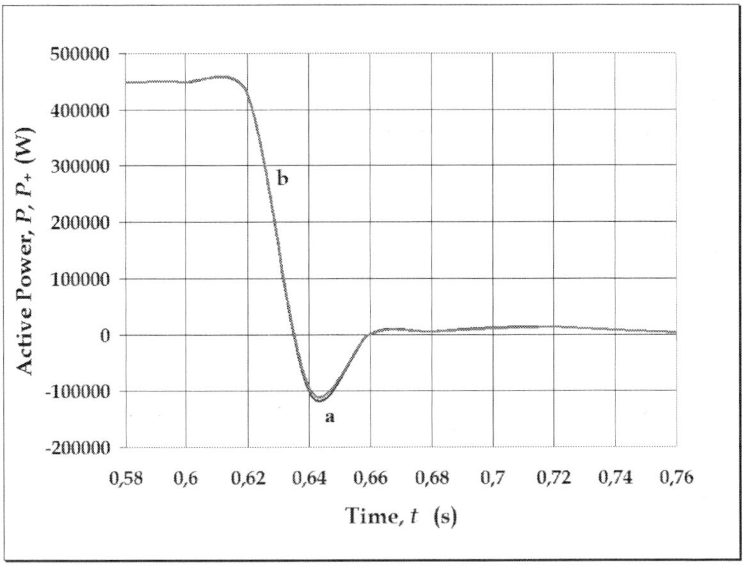

Fig. 14. Active powers: (a) Traditional, (b) Unified Theory

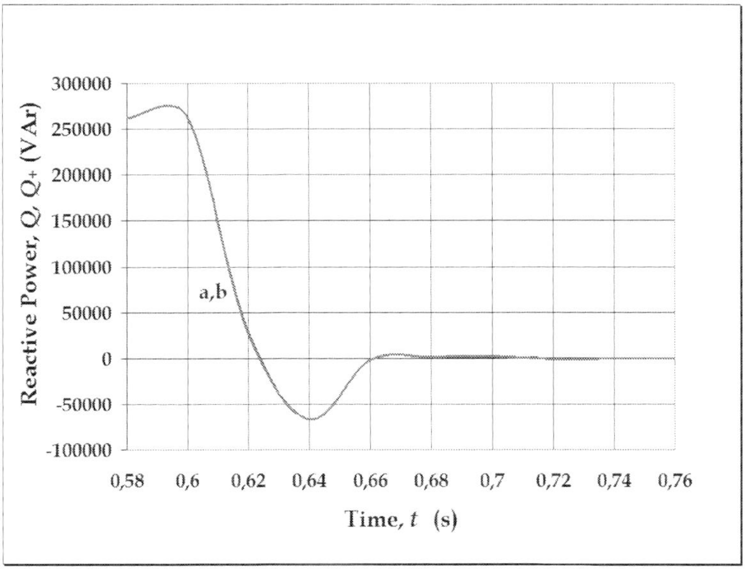

Fig. 15. Reactive powers: (a) traditional, (b) Unified Theory

Spanish and German grid code requirements was verified by the wind farm in presence of the analyzed two-phase dip whether the Unified Theory is used. However, the application of the traditional theory is very complicated since the traditional active and reactive currents have different sign and value in each grid phases (figs. 16 and 18) and traditional active and reactive powers contain negative-sequence components. Unified Theory's positive-sequence active and reactive currents verify grid code requirements because their values are not increased during the fault (figs. 17a and 19a). Moreover, the maintenance of the positive-sequence reactive power is explained by an important consumption of the positive-sequence reactive current caused by the unbalances (fig. 19c), which compensate the increasing of the reactive current demanded by the grid (fig. 19b). Figure 20 shows how the duration of positive-sequence active power consumptions is less than the time period of the traditional active power consumptions and, thus, the accomplishment of the grid code requirements is improved. This fact occurs because a short positive-sequence active power delivery caused by the unbalances (fig. 21b). Difference between the traditional and the Unified Theory's reactive powers (fig. 22) defines the negative-sequence component of the reactive power which originates reverse magnetic fields and causes wind-generator malfunction. Positive-sequence reactive power is decreased by a strong reactive power consumption caused by the unbalances during the voltage dip (fig. 23b). This reduction of the positive-sequence reactive current supplied to the grid is convenient for the accomplishment of the grid code requirements.

The analysis of the two-phase voltage dip shows the Unified Theory is clearly better than the traditional theory for verifying the accomplishment of the grid code requirements, since that theory uses quantities more related with the active and reactive phenomena and it gives up additional information about those phenomena.

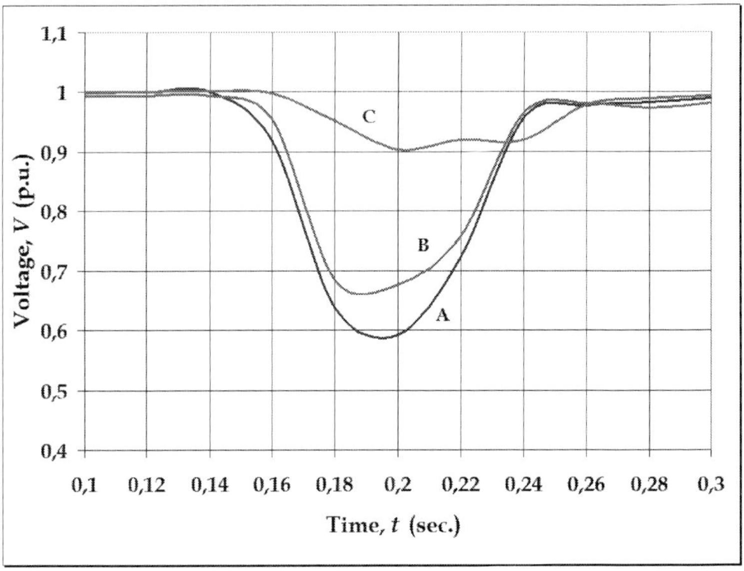

Fig. 16. Two-phase voltage dip

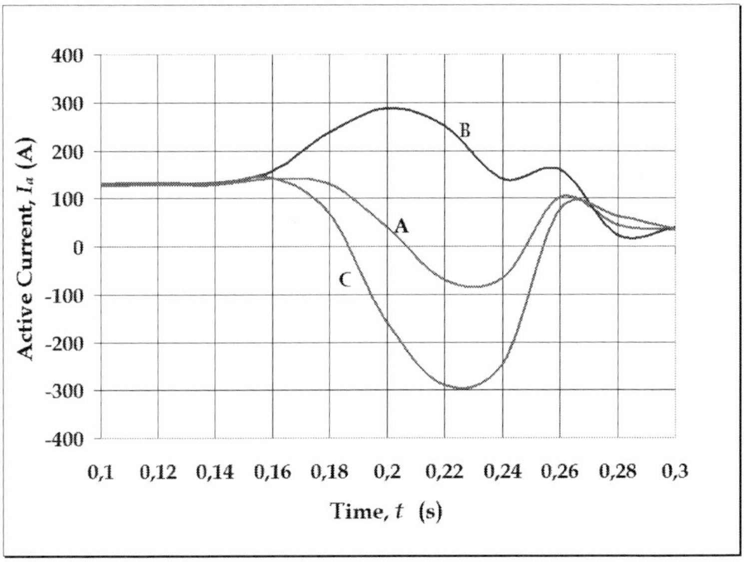

Fig. 17. Phase active currents

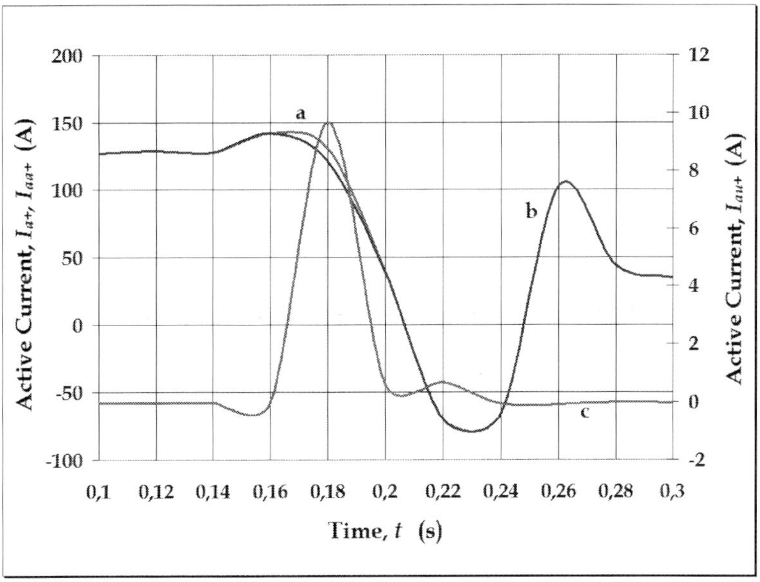

Fig. 18. Unified Theory's active currents: (a) total, (b) due to the active loads, (c) caused by the unbalances

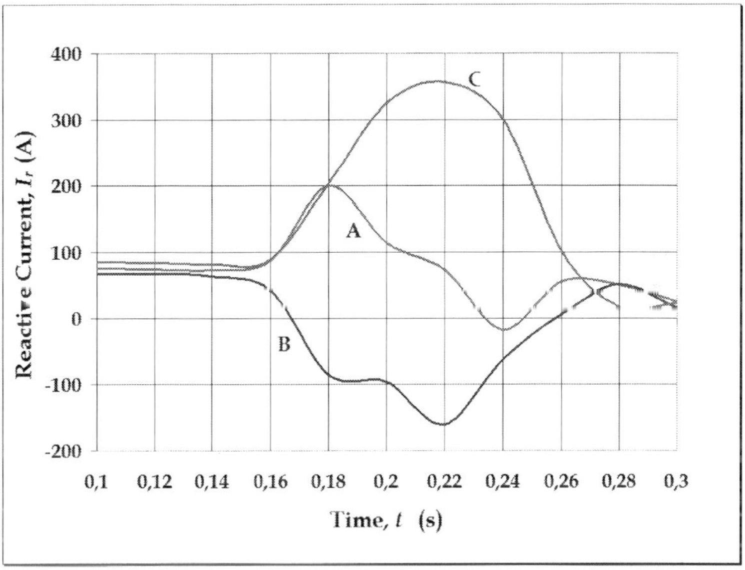

Fig. 19. Phase reactive currents

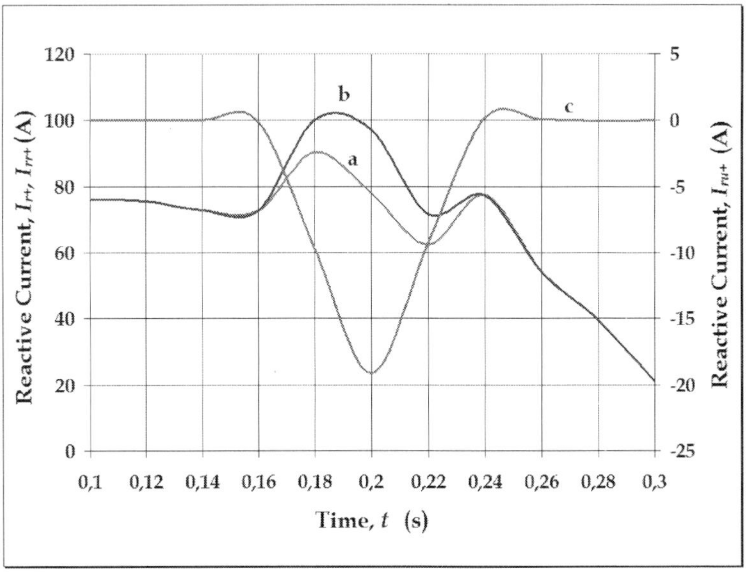

Fig. 20. Unified Theory's reactive currents: (a) total, (b) due to the reactive loads, (c) caused by the unbalances

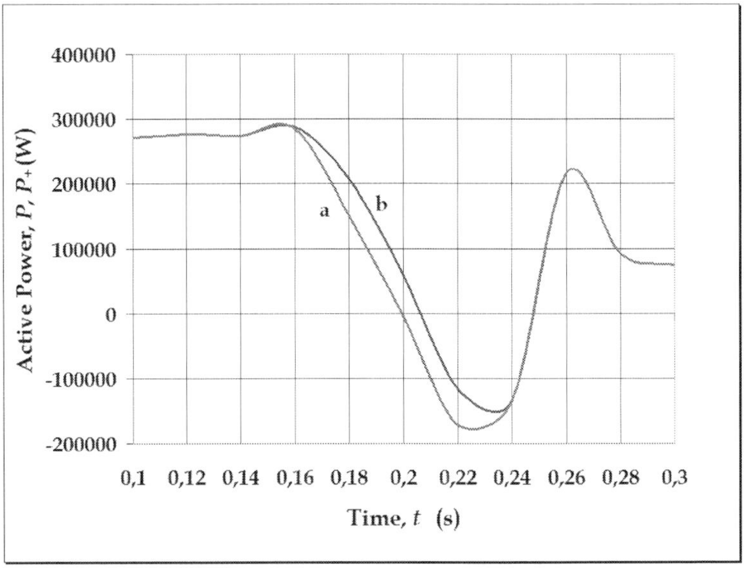

Fig. 21. Active powers: (a) traditional theory, (b) Unified Theory

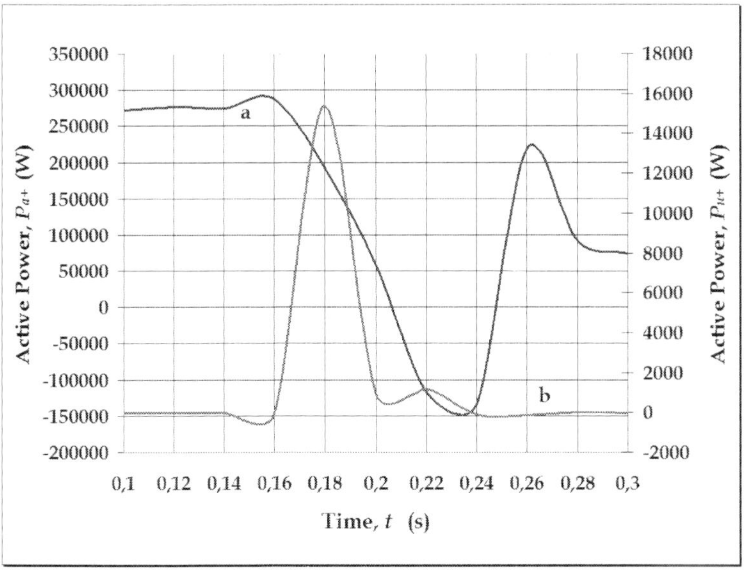

Fig. 22. Unified Theory's active powers components: (a) due to the active loads, (b) caused by the unbalances

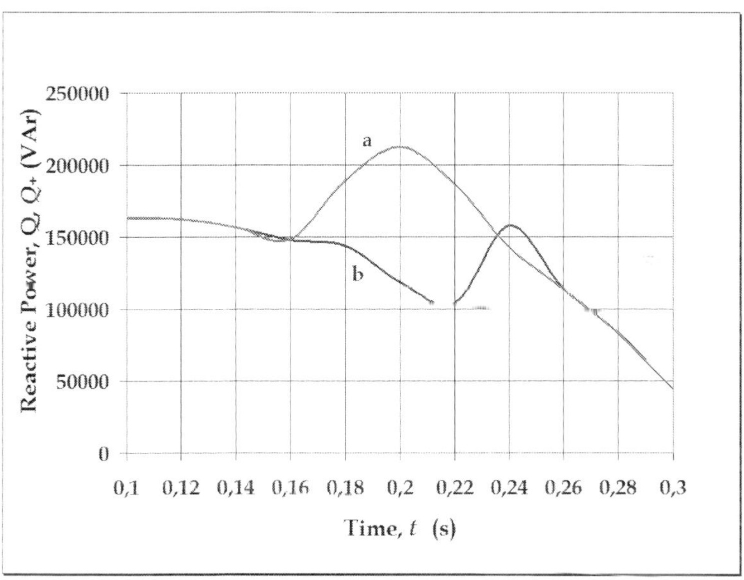

Fig. 23. Reactive powers: (a) traditional theory, (b) Unified Theory

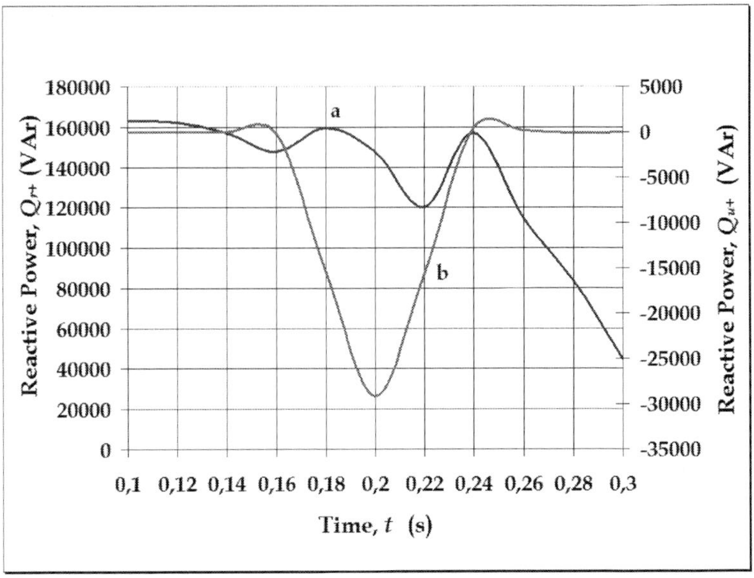

Fig. 24. Unified Theory's reactive power components: (a) due to the reactive loads, (b) caused by the unbalances

5. Conclusions

The Spanish Grid Code and the grid codes from other countries require some quantities, such as active and reactive currents and powers, must be controlled in order to avoid unexpected disconnections of the wind farms submitted to voltage dips. These grid codes implicitly propose the traditional well-known formulations, included in the IEEE Standard 1459-2010, for measuring active and reactive powers and currents. For balanced voltage dips, these formulations are adequate to verify grid code requirements, although the different values of the active and reactive phase currents may difficult the verification process. However, for unbalanced voltage dips, traditional formulations include components which are a result of the imbalances and, thus, mistakes in the magnitude and duration of the active and reactive quantities may be presented.

Fundamental positive-sequence active and reactive formulations, also included in the IEEE Standard 1459-2010, are a more adequate alternative than the traditional theory for verifying the accomplishment of the grid code requirements. Several reasons justify the use of the fundamental positive-sequence quantities: (a) active and reactive currents have only one component so much for balanced as unbalanced voltage dips and, thus, the verification process of the grid code requirements is simplified; (b) positive-sequence active and reactive powers do not contain negative-sequence components caused by the voltage unbalances and, thus, these quantities exactly quantify active and reactive phenomena effects, respectively; (c) positive-sequence active and reactive powers and currents can be decomposed into two components, due to the loads and caused by the unbalances.

This decomposition established by the Unified Theory has been expressed in section 2. It shows how imbalances of supplies and loads originate additional positive-sequence powers and currents, which either can increase or decrease total values of these quantities and, therefore, the accomplishment of the grid code requirements can be better explained and new wind-generator support procedures can be proposed by applying the Unified Theory.

6. References

Emmanuel, A.E. (1999). Apparent Power Definitions for Three-Phase Systems. *IEEE Transactions on Power Delivery*, Vol.10, No.3, July, 1999, 767-772, ISSN 0885-8977.

E.ON Netz. (2006). *Grid Code: High and extra high voltage*. E.ON Netz GmbH, Bayreuth (Germany), April, 2006.

Kim, H., Blaabjerg, F. & Bak-Jensen, B. (2002). Spectral Analysis of Instantaneous Powers in Single-Phase and Three-Phase Systems with Use of p-q-r Theory. *IEEE Transactions on Power Electronics*, Vol.17, No.5, September, 2002, 711-720, ISSN 0885-8993.

Industry, Tourism and Commerce Spanish Ministry. (2006). *Operation Procedure O.P. 12.: Response requirements in front of voltage dip at wind farms utilities*. BOE 254, 37017-37019, October, 2006, Madrid.

León, V., Montañana, J., Roger, J., Gómez, E., Cañas, M., Fuentes, J.A. & Molina, A. (2009). Verification of the Reactive Power Requirements in Wind Farms. *Proceedings of IEEE PowerTech 2009*, ISBN 978-1-4244-2234-0, Bucharest, June-July, 2009.

León, V., Montañana, J., Roger, J., Gómez, E., Cañas, M., Fuentes, J.A. & Molina, A. (2009). Reactive power and current formulations for wind farms Spanish grid code. *Proceedings of EEM 2009*, ISBN 978-1-4244-4455-7, Leuven, May, 2009.

León, V., Montañana, J., Roger, J., Gómez, E., Cañas, M., Fuentes, J.A. & Molina, A. (2009). Estimation of Wind Farms Working in Presence of Voltage Dips Using the IEEE Std. 1459-2000. *Proceedings of PSCE'09*, ISBN 978-1-4244-3810-5, Seattle, March, 2009.

León-Martínez, V., Montañana-Romeu, J. (2009). Method and system for calculating the reactive power in disturbed three-phase networks. *PCT/ES 2009/000370*, July, 2009.

León-Martínez, V., Montañana-Romeu, J., Giner-García, J., Cazorla-Navarro, A., Roger-Folch, J. (2007). Power Quality Effects on the Measurement of Reactive Power in Three-Phase Power Systems in the Light of the IEEE Standard 1459-2000. *Proceedings of EPQU 2007*, ISBN 978-84-690-9441-9, Barcelona, October, 2007.

León-Martínez, V., Giner-García, J., Montañana-Romeu, J. & Cazorla-Navarro, A. (2001). Efficiency in electrical installations. New power definitions. *Mundo Electrónico*. No.322, July, 2001, 28-32. ISSN 0300-3787.

Power System Instrumentation & Measurement Committee. (2010). *IEEE Std. 1459-2010, IEEE Standard Definitions for the Measurement of Electric Power Quantities Under Sinusoidal, Non-Sinusoidal, Balanced or Unbalanced Conditions*, The Institute of

Electrical and Electronics Engineers, March, 2010, ISBN 978-0-7381-6058-0, New York.

Spanish Wind Enegy Association. (2008). *Offprint of the Operation Procedure O.P. 12.2: Technical requirements for wind power and photovoltaic installations and any generating facilities whose technology does not consist on a synchronous generator directly connected to the grid. Utilities connected to the transport grid and generating equipement: minimum design requirements, equipment, operation, deployment and security*. www.aeeolica.es.

Part 3

Empirical Approaches to Estimating Hydraulic Conductivity

4

Frequency Control of Isolated Power System with Wind Farm by Using Flywheel Energy Storage System

Rion Takahashi
Kitami Institute of Technology
Japan

1. Introduction

For the recent expansion of renewable energy applications, wind energy generation is receiving much interest all over the world. Many large wind farms have been installed so far and recently huge offshore wind farms have also been installed. However, the frequency variation of power system due to wind generator output fluctuations is a serious problem. If installations of wind farms continue to increase, frequency control of power system by the main sources, that is, hydraulic and thermal power stations, will be difficult in the near future, especially in an isolated power system like a small island which has weak capability of power regulation. In such a case, the installation may be restricted even though it is a small wind farm. Though there is such a difficulty, an introduction of the wind energy utilization is much effective in an isolated power system, because main power plant in a small island is mostly a diesel engine driven generating plant and it has no good effect on the environment. Hence, some strategies are necessary to improve the stability of wind farm output. According to such situations, an application of battery system for the output power smoothing has been investigated so far, and some experimental studies using practical facilities are being performed. The battery system is suitable for power compensation with relatively long period like load leveling. However, since rapid response is necessary to compensate power variations in an isolated power system, the battery system may not be appropriate because charging or discharging speed of the battery is not so fast due to its chemical process. Moreover, the same capacity of electronic power converter as that of the battery power rating is required. In addition life time of battery is, in general, not so long and thus frequent replacement of battery cell will be needed. These characteristics cause cost increase. On the other hand, the application of Flywheel Energy Storage System (called 'FESS' hereinafter) for power compensation is very effective. This system has characteristics of large energy storage capacity, long life, and rapid response of power control. It has a heavy weight rotating mass connected to an adjustable speed generator. This chapter adopts an adjustable speed generator with secondary AC excitation as a driving machine of rotating mass, because this type of generator has already been put into practice in pumped storage hydro power plants in Japan [1]. There are also some practical applications of FESS to improve power system stability [2]. The adjustable speed generators with secondary AC excitation can control not only active power output but also reactive power output rapidly

and independently. Thus smoothing of both output power and grid voltage fluctuations in wind farm is possible by installing FESS with the adjustable speed generator. In addition, since only small capacity of electronic power converter is needed in this system, the total cost can be decreased. Therefore, the FESS can be effective on smoothing of wind farm output fluctuation, resulting in the frequency stabilization of the power system.
With these points as background, this chapter proposes a control strategy of FESS to reduce the frequency variation in an isolated power system including a wind farm. The main features are as follows: 1) Cooperation with the main power plant, i.e., output of the main power plant is adjusted in co-operation with the FESS depending on its energy charge level; 2) Direct frequency control. In the case of large power system, generally, smoothing of rapid change of wind farm output in short period is performed by energy storage system, while slow change in long term is absorbed by other power plants for frequency control. However in the isolated power system, single or a few main source generators can hardly regulate slow power fluctuation. Therefore direct frequency control by energy storage system is desirable. In order to evaluate the effectiveness of the proposed method, computer simulation analyses are performed by using PSCAD/EMTDC [3].

2. Example of model system

Overview of FESS operation

Fig. 1 shows an overview of FESS operation proposed in this chapter. The isolated power system consists of main power supply, a consumer load and a wind farm. FESS is installed near the wind farm. FESS detects the network frequency and stabilizes it by supplying or absorbing active power to/from the network. FESS also sends a command to the main power supply to adjust its output so as to keep suitable stored energy level of FESS.

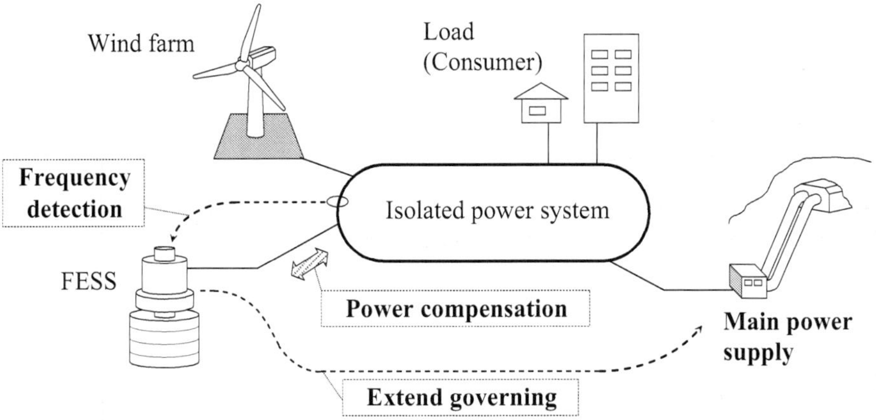

Fig. 1. Overview of FESS operation

Brief configuration of power system

Fig. 2 shows the power system model used in this chapter. A Wind Farm (WF) is modeled by a single induction generator with a wind turbine operating almost at constant speed. The FESS is installed to the grid point of wind farm. A Synchronous Generator (SG) as a main source generator which is driven by a diesel engine is connected to the grid point through a transmission line, and resistive loads are connected to the both ends of the line.

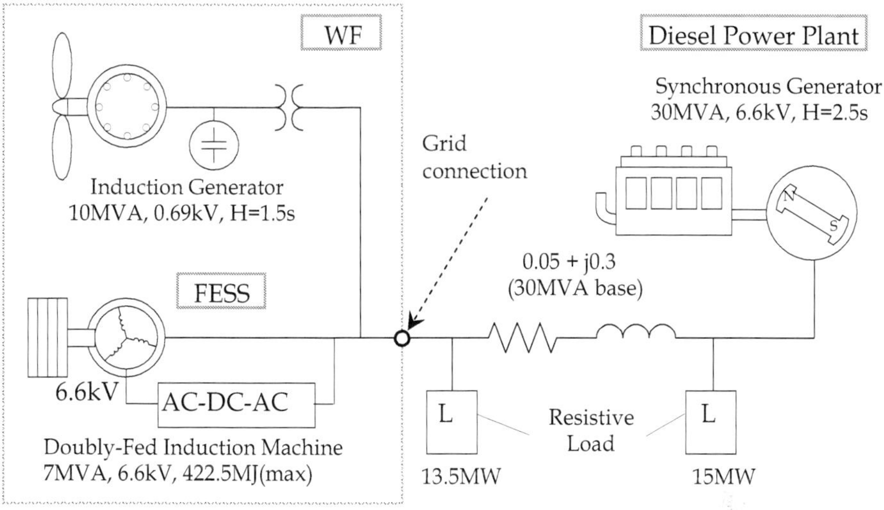

Fig. 2. Model system of an isolated power system

Configuration of FESS

Fig. 3 shows a model configuration of FESS. The FESS consists of the adjustable speed generator, the flywheel mass for kinetic energy storage, and secondary excitation circuit for adjustable speed control [4]. The adjustable speed generator has basically the same construction as that of a wound rotor induction machine. The secondary excitation power is supplied from the terminal of FESS, and converted to DC power by the converter, then again converted to low frequency AC power by the inverter and supplied to the rotor. Thus, the rotor can rotate at asynchronous speed. The inverter controls active and reactive power output (P_T and Q_T) of the generator, and the converter controls DC link voltage E_{DC} and reactive power Q_L flowing into the secondary excitation circuit. These electronic power converters are modeled as 6 force-commutated power switches connected in a bridge configuration as shown in Fig. 4. A sinusoidal PWM operation is carried out and switching signals are generated by applying triangular carrier wave comparison. Conventional PI controllers are used for the inverter and the converter control as shown in Fig. 5 and 6 respectively. Parameters of the FESS generator are shown in Table II.

A method of frequency stabilization by using FESS

The main purpose of this study is to reduce the network frequency variation by using FESS. The configuration of the control system for the frequency stabilization is shown in Fig. 7.
Reference of active power output of FESS, $P_{T(ref)}$, is determined according to the deviation of network frequency, which is detected by PLL at the terminal of FESS. When the frequency is decreased, FESS supplies active power to the network. When the frequency is increased, FESS absorbs active power from the network. These control schemes correspond to block (A) in Fig. 7. At the same time, $P_{T(ref)}$ is modified to prevent a shortage or an excess of the

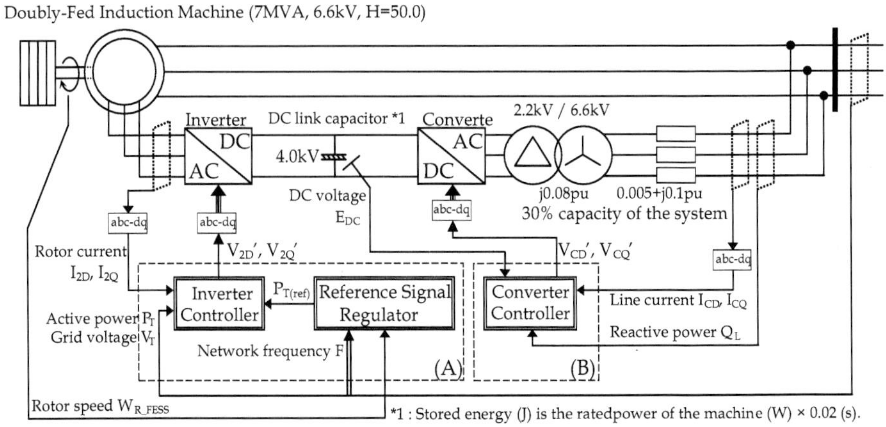

Fig. 3. FESS circuit configuration

stored energy of FESS. In this study, the maximum and the minimum rotor speeds of FESS are specified 130% (1.3pu) and 70% (0.7pu) of the rated speed respectively. Considering these boundary speeds, the value of $P_{T(ref)}$ is modified to a lower (or a higher) value when the rotor speed is under (or over) 1.044pu, at which the stored energy becomes a half of the maximum storage energy. These control schemes correspond to block (B) in Fig. 7. Fig. 7 also includes a rule of FESS control to avoid operating under 0.7pu or over 1.3pu rotor speed as shown in Table I.

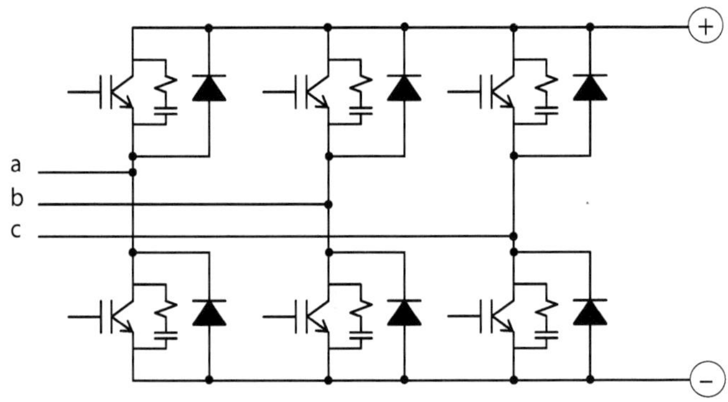

Fig. 4. Model of power converter

			Frequency < 50	Frequency > 50
	W_{R_FESS}	> 1.3	1	0
1.3 >	W_{R_FESS}	> 0.7	1	1
0.7 >	W_{R_FESS}		0	1

Table I. Rule of FESS control

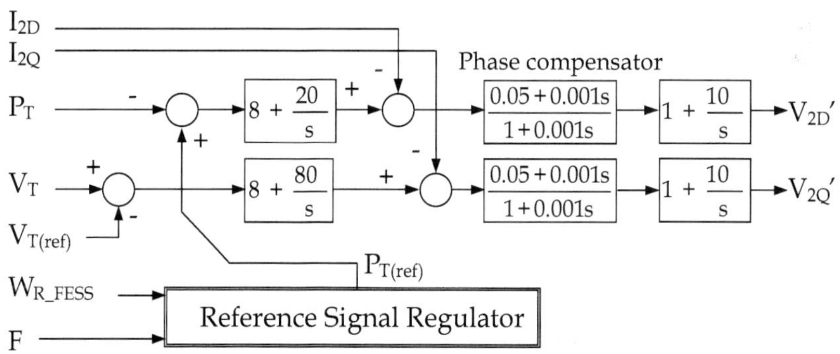

Fig. 5. Output power controller of FESS

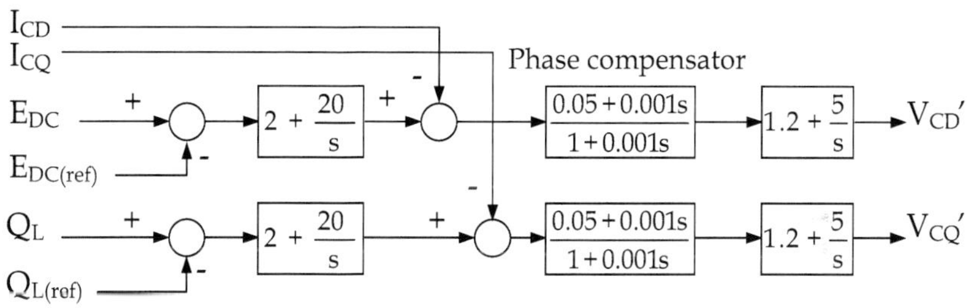

Fig. 6. Excitation power controller of FESS

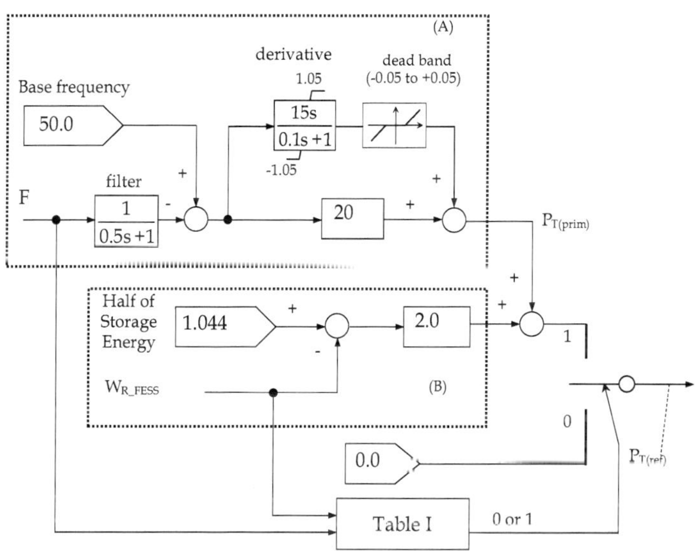

Fig. 7. Reference signal controller for frequency stabilization

Wind farm model

The wind farm consists of an induction generator and a wind turbine. An aerodynamic characteristic of the turbine blade expressed by eqs.(2) and (3) is adopted [5]. The captured power is expressed by eq.(1). Since the induction generator is operated at almost constant speed (approx. 1.0 to 1.01 pu), the output power changes widely with respect to wind speed variations. Generally, a wind turbine is equipped with a pitch angle controller. The conventional pitch controller shown in Fig. 8, that maintains the output of the generator to be the rated power when the wind speed is over the rated speed, is also considered in this study. Parameters of the wind generator (IG) are shown in Table II.

$$P_M = \frac{1}{2}\rho C_p(\lambda)\pi R^2 V_W^3 [W] \tag{1}$$

$$C_p(\lambda) = 0.5(\Gamma - 0.0.2\beta^2 - 5.6)e^{-0.17\Gamma} \tag{2}$$

$$\Gamma = \frac{R}{\lambda} \cdot \frac{3600}{1609} \tag{3}$$

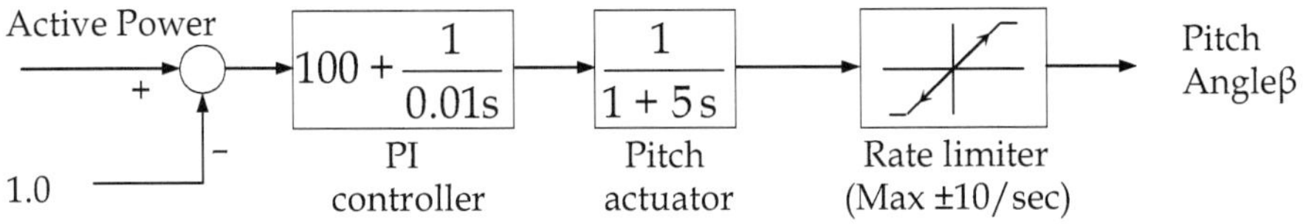

Fig. 8. Pitch angle controller of wind turbine

	IG	FESS
Stator resistance (pu)	0.01	0.02
Stator leakage reactance (pu)	0.07	0.08
Magnetizing reactance (pu)	4.1	3.5
Rotor resistance (pu)	0.007	0.02
Rotor leakage reactance (pu)	0.07	0.08

Table II. Parameters of induction machines.

Synchronous generator model

A Synchronous Generator (SG) is considered as a main power supply unit in the network in this study, which is assumed to be a diesel engine driven power plant. The characteristics of the diesel engine and its governor system in [6] are considered. The governor controls fuel supply to maintain the engine speed at the synchronous speed. Its block diagram is shown in Fig. 9, and its parameters are shown in Table III.

If FESS regulates the network frequency by its power compensation, the output of SG may not change, because the network frequency is controlled to be constant. Consequently, there is a possibility that FESS performs all of the network frequency control instead of SG. In such case, when the stored energy in FESS becomes full or empty, the power balance of the network cannot be maintained and thus the network frequency can deviate significantly. To avoid such situation, the output of SG also needs to be regulated according to the stored energy of FESS. In this chapter, a cooperative control is proposed, in which the output of SG is increased (or decreased) when the rotor speed of FESS is below (or over) 1.044pu which corresponds to a half of the maximum storage energy of FESS. But if the additional command to the main source generator changes fast, its output will also vary widely, and then it suffers large mechanical stress. Therefore a control gain is set for the additional command to change slowly as shown in Fig. 10. The governor of SG in this study has been designed to control only engine speed, and thus the output of SG can be changed by modifying a monitored signal of the engine speed to the governor. These control systems are shown in Fig. 10. In addition, a simple AVR model shown in Fig. 11 is used in SG model. Parameters of the synchronous generator (SG) are shown in Table IV.

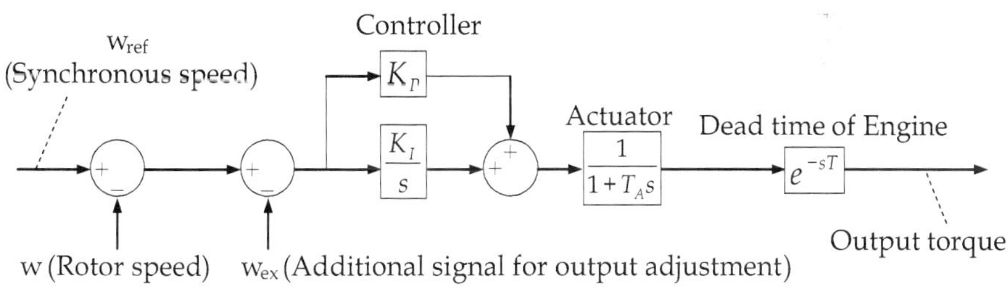

Fig. 9. Governor model of the diesel engine

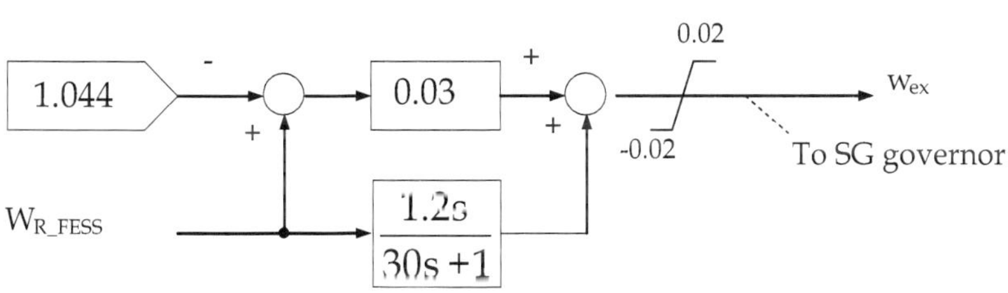

Fig. 10. Additional signal controller for output adjustment of the diesel engine

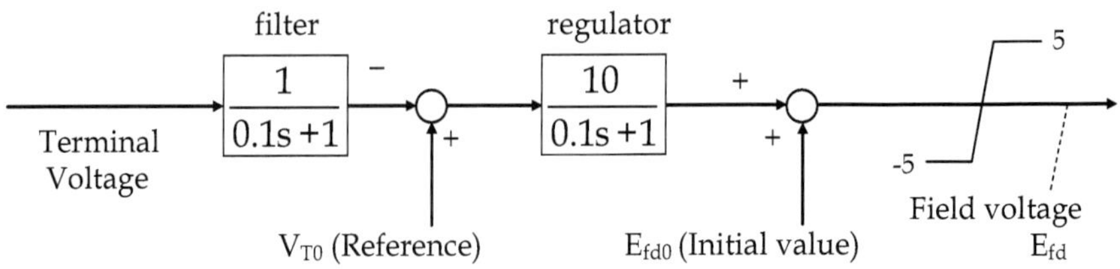

Fig. 11. AVR model of the synchronous generator

Proportional Gain of K_P	8.0
Integral Gain K_I	2.0
Pilot servo time constant T_A	0.2 s
Dead time of engine T	0.25 s

Table III. Parameters of the diesel engine governor

Armature resistance (pu)	0.0025	
Stator leakage reactance (pu)	0.14	
Field resistance (pu)	0.0004	
Field leakage reactance (pu)	0.2	
	D-axis	Q-axis
Magnetizing reactance (pu)	1.66	0.91
Damper resistance (pu)	0.005	0.0084
Damper leakage reactance (pu)	0.044	0.106

Table IV. Parameters of synchronous generator.

3. Simulation example

A. Condition

A determination of the energy storage capacity is very important for designing energy storage system. In this chapter, the energy storage capacity of FESS is determined from a point of view of adequate frequency control ability but reducing it as small as possible. The power rating of FESS is decided as 70% of that of the wind farm since instantaneous output change of the wind farm can hardly reach its power rating in normal operation.
Comparative study between the proposed frequency control method (shown in Fig. 7 and Table I) and a power smoothing method (shown in Fig. 12 and Table V) which is generally considered in a wind farm connected to large power system, is performed in the simulation analysis here. In conventional power smoothing method, an energy storage system only smoothes wind farm output fluctuations, and slow change of wind farm output is absorbed by several thermal and hydraulic power plants installed as main generators in large power system. However, since the total power rating and the number of main power generators are limited in the case of an isolated power system, power regulation may become difficult even when wind farm output fluctuation is small. Moreover, the output of main power generators should be adjusted also to maintain the amount of residual energy of storage system. If the stored energy is not regulated suitably, power balance of the isolated power system cannot be kept when the stored energy reaches full or empty level. Therefore, it can be said that the frequency stabilization in the case of an isolated power system cannot be achieved only by the conventional power smoothing scheme.

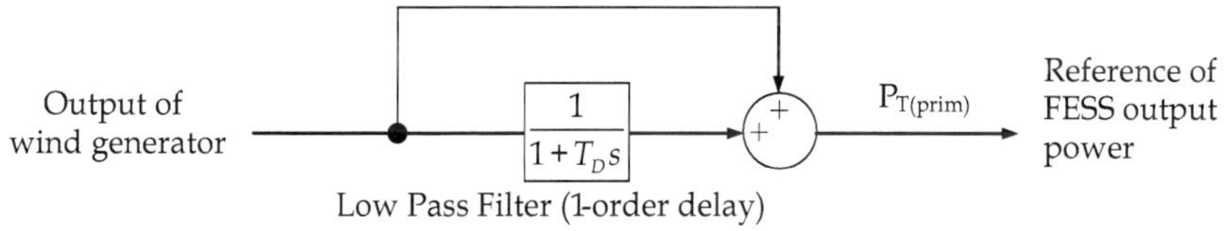

Fig. 12. Reference signal regulator of the FESS for power smoothing control

			$P_{ref} < 0$	$P_{ref} > 0$
	W_{R_FESS}	> 1.3	1	0
1.3 >	W_{R_FESS}	> 0.7	1	1
0.7 >	W_{R_FESS}		0	1

Table V. Rule of FESS control for power smoothing

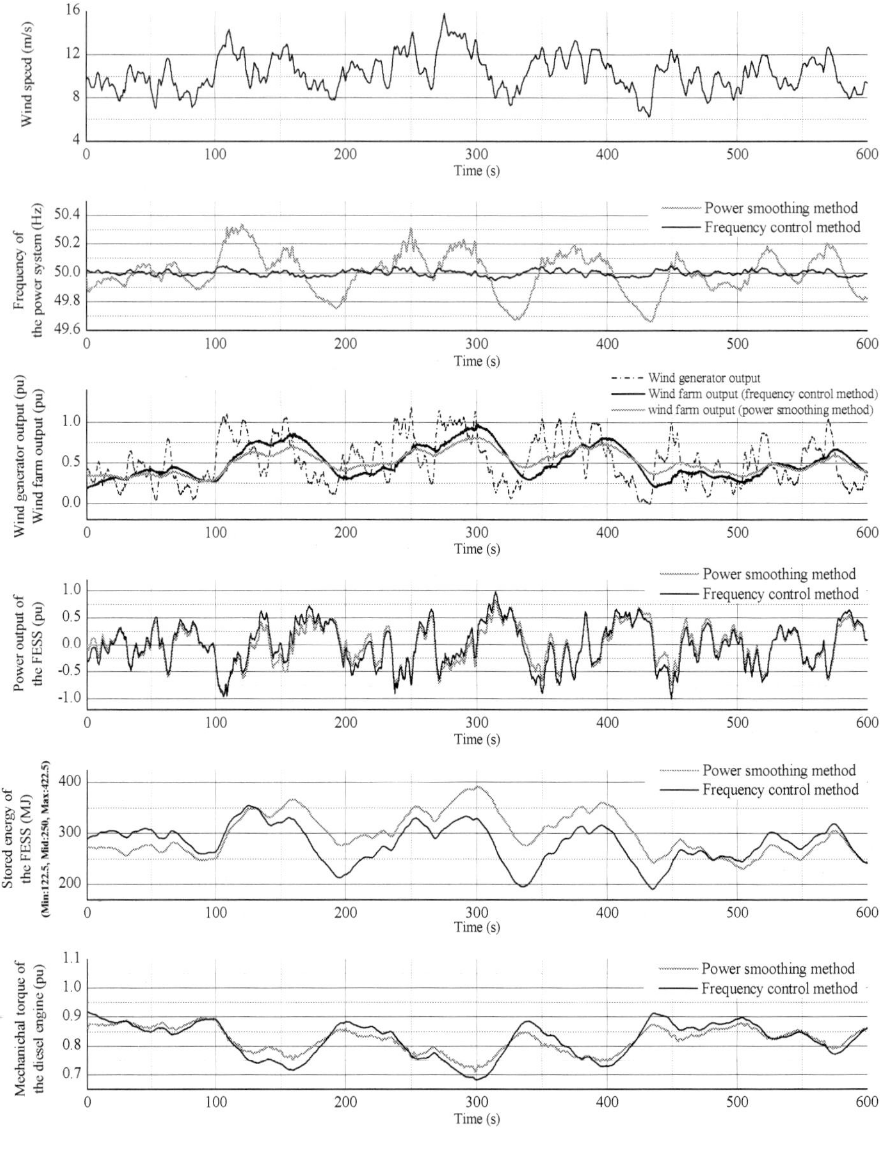

Fig. 13. System response under the frequency control method and power smoothing method

In the comparative study, simulations by the conventional output power smoothing method have also been performed, in which the reference output from the WF to the grid is determined by inputting the net WF output into a first order delay transfer function and FESS supply the difference between the reference power and the net output to follow the WF-to-grid output to the reference value as shown in Fig. 12. Therefore block A in Fig. 7 is replaced by Fig. 12 and the rule of FESS control shown in Table I is also replaced by Table V. In the conventional method, the cooperation control with the main power plant is impossible. The power rating of FESS is chosen to be 7MVA, same as that in the frequency control method, and the time constant of the first order delay is set to 30s.

B. Results

Fig. 13 shows system responses under the proposed frequency control method and the conventional power smoothing method. The frequency deviation reaches about 0.3Hz at the maximum in the case of the power smoothing method, but it is regulated within about 0.05Hz in the case of the frequency control method. The stored energy of FESS is remained well between the maximum and the minimum levels, from which there may be a possibility that the energy storage capacity of FESS can be reduced. Responses of the prime mover output (diesel engine output) are almost the same in both methods.

4. Summary

This chapter has proposed a new method of network frequency regulation by using Flywheel Energy Storage System (FESS) for an isolated power system including a wind farm, and the validity of the proposed method has been evaluated by computer simulations. From the comparative study between the proposed method and the conventional output smoothing control of wind farm, it has been shown that the proposed method is very effective on the stabilization of network frequency in an isolated small power system.

The proposed method can be applied basically not only to a FESS system but also other types of energy storage system. Therefore the proposed method can contribute to expand wind energy utilization into isolated power systems like a small island.

5. References

[1] T. Kuwabara, A. Shibuya, H. Furuta, E. Kita, and K. Mitsuhashi : "Design and Dynamic Response Characteristics of 400 MW Adjustable Speed Pumped Storage Unit for Ohkawachi Power Station," IEEE Transactions on Energy Conversion, Vol. 11, No. 2, pp. 376-384, June 1996.

[2] M. Kazuma, U. Yuuetsu : "Hydroelectric Power Technologies Contributing to Power System Quality Improvement", TOSHIBA REVIEW, VOL.58, NO.7, 2003.

[3] Manitoba HVDC Research Centre (http://www.hvdc.ca/)

[4] R.Takahashi, J.Tamura, Y.Tada, A.Kurita: "Model Derivation of an Adjustable Speed Generator and Its Excitation Control System", Proc. of 14-th Power System Computation Conference, Session-06, paper-4, June 2002.

[5] O. Wasynczuk, D. T. Man, J. P. Sullivan : "Dynamic Behavior of a Class of Wind Turbine Generator During Random Wind Fluctuations", Trans. of IEEE on Power Apparatus and Systems, Vol. PAS-100, No.6, pp.2873-2845, June 1981.

[6] Sanjoy Roy, O.P.Malik, G.S.Hope : "A k-Step Predictive Scheme for Speed Control of Diesel Driven Power Plants", IEEE Transactions on Industry Applications, Vol. 29, No. 2, pp. 389-396, March/April 1993.

5

Control Scheme of Hybrid Wind-Diesel Power Generation System

Cuk Supriyadi A.N[1], Takuhei Hashiguchi[1], Tadahiro Goda[1] and Tumiran[2]
[1]*Graduate School of Information Science and Electrical Engineering, Kyushu University, Fukuoka, 819-0395*
[2]*Department of Electrical and Information Technology Faculty of Engineering, Gadjah Mada University, Yogyakarta,55281*
[1]*Japan*
[2]*Indonesia*

1. Introduction

Global warming is one of the most serious enviromental problems facing the world community today. It is typified by increasing the average temperature of Earth's surface and extremes of weather both hot and cold. Therefore, implementing a smart and renewable energies such as wind power, photo voltaic etc are expected to deeply reduce heat-trapping emissions. Moreover, wind power is expected to be economically attractive when the wind speed of the proposed site is considerable for electrical generation and electric energy is not easily available from the grid (Ackermann, 2005). This situation is usually found on islands and/or in remote localities. However, wind power is intermittent due to worst case weather conditions such as an extended period of overcast skies or when there is no wind for several weeks. As a result, wind power generation is variable and unpredictable.

The hybrid wind power with diesel generation has been suggested (Hunter, 1994) and (Lipman, 1989) to handle the problem above. A hybrid wind diesel system is very reliable because the diesel acts as a cushion to take care of variation in wind speed and would always maintain an average power equal to the set point. However, in addition to the unsteady nature of wind, another serious problem faced by the isolated power generation is the frequent change in load demands. This may cause large and severe oscillation of power. The fluctuation of output power of such renewable sources may cause a serious problem of frequency and voltage fluctuation of the grid, especially, in the case of isolated microgrid, which is the a small power supply network consisting of some renewable sources and loads. In the worst case, the system may lose stability if the system frequency can not be maintained in the acceptable range.

Control schemes to enhance stability in a hybrid wind - diesel power system have been proposed by much researchers in the previous work. The programmed pitch controller (PPC) in the wind side can be expected to be a cost-effective device for reducing frequency deviation (Bhatti et. al ,1997) and (Das et. al, 1999). Nevertheless, under the sudden change of load demands and random wind power input, the pitch controller of the wind side and the governor of the diesel side may no longer be able to effectively control the system frequency due to theirs slow response. To overcome this problem, an Energy Storage (ES), which is able

to supply and absorb active power rapidly, has been highly expected as one of the most effective controller of system frequency (Tripathy et. al. 1997) and (Tripathy et. al. 1997).
In this chapter, Superconducting Magnetic Energy Storage (SMES) is used as Energy Storage. It is able to compensate the fluctuation of wind power generation. The SMES unit is a device that stores energy in the magnetic field generated by the direct currents flowing through a superconducting coil. Since energy is stored as a circulating current, energy can be drawn from the SMES unit with almost instantaneous response with energy stored or delivered over periods ranging from a fraction of a second to several hours (Ribeiro et.al, 2001). Because direct current flows with negligible losses in superconductors, the SMES unit can be used for small and large scale energy storage and rapid charge/discharge applications. The SMES system consists of a large superconducting coil at the cryogenic temperature. The coil is kept at cryogenic (superconductive) temperature by a refrigeration system designed to meet the superconducting properties of the special materials used to fabricate the magnetic coil. A power conversion/conditioning system connects the SMES unit to an ac power system, which has an inverter that converts the dc output of the storage device to ac during discharge and the ac to dc for recharging the storage device (Schainker, 2004).
The SMES systems have several advantages. The SMES coil has the ability to release large quantities of power within a fraction of a cycle, and then fully recharge in just minutes. The SMES unit can store and discharge DC power at efficiencies of 98% or more and switch between charging and discharging within 17 milliseconds. This quick, high-power response is very efficient and economical. The SMES manufacturers cite controllability, reliability and no degradation in performance over the life of the system as prime advantages of SMES systems. The estimated life of a typical system is at least 20 years (Schainker, 2004).
In power system, the SMES is capable of supplying both active and reactive powers simultaneously and quickly. Thus, it is able to enhance the power system stability and reliability dramatically (Jiang & Chu, 2001) and (Simo& Kamwa, 1995). Primarily, the SMES unit was aimed to store energy during the off-peak load period and release it in the peak load period. It has been shown that the SMES is able to supply the active and reactive power simultaneously and damp the oscillations in an power system (Simo& Kamwa, 1995) and (Wu & Lee, 1993). In fact, the SMES can also be used as a PSS, if the control scheme is suitably designed (maschowski & Nelles, 1992). Besides, the applications of the SMES also include load regulation, transmission stabilization, uninterruptible power supply, power compensation, voltage control and improving customer power quality, etc. (Buckles & Hassenzahl, 2000). Moreover, the SMES also has been successfully applied to solve many problems in power systems such as an improvement of power system dynamics (Rabbani et.al., 1998) and (Devotta & Rabbani,2000), a frequency control in interconnected power systems (tripathy,1997) and (Ngamroo,2005), an improvement of power quality (Chu et.al. 2001), a stabilization of sub-synchronous oscillation in the turbine-generator (Devotta et.al. 1999), a load leveling (Abdelsalam et.al. 1987) etc.
Several design methods to design SMES have been successfully proposed, such as a proportional control (Banerjee et.al. 1990), a digital control (Tripathy & Juengst, 1997), an adaptive control (Tripathy et.al. 1997), a neural network (Demiroren et.al. 2003) and a fuzzy control (Demiroren & Yesil 2004), etc. Despite the potential of modern control techniques with different structures, power system utilities still prefer the fixed structure controller. The reasons behind that might be the ease of on-line tuning and the lack of the assurance of stability related to some adaptive or variable structure techniques. On the other hand, various generating and loading conditions, wind power fluctuations, variation of system

parameters and system nonlinearities etc., result in system uncertainties. The SMES controllers in these works have been designed without considering system uncertainties. The robust stability of resulted SMES controllers against uncertainties cannot be guaranteed. They may fail to operate and stabilize the power system.

To enhance the robustness, many research works have been successfully applied robust control theories to design of PSS and damping controllers of flexible AC transmission systems (FACTS) devices. In (Djukanovic et.al. 1999) and (Yu et.al. 2001), the structured singular value has been applied to design robust PSS and static var compensator (SVC), respectively. In (Zhu et.al. 2003) and (Rahim & Kandlawala, 2004), the H_∞ control approach has been used to design robust PSS and FACTS devices. The presented robust controllers above provide satisfactory effects on damping of power system oscillations. Nevertheless, selection of weighting functions becomes an inevitable problem that is difficult to solve. Furthermore, an order of designed controller depends on that of the system. This leads to the complex structure controllers. In (wang et.al. 2002) and (Tan & wang, 2004), the robust non-linear control based on a direct feedback linearization technique has been applied to design an excitation system, a thyristor controlled series capacitor (TCSC) and a SMES. However, the drawback of this design method is a tuning of Q and R matrices for solving Riccati equation by trial and error. Besides, the resulted controllers are established by a state feedback scheme which is not easy to implement in practical systems.

This chapter presents a controller design of programmed pitch controller (PPC) and Energy storage (ES) to control frequency oscillation in a hybrid wind-diesel power generation. To take system uncertainties into account in the control design, the inverse additive perturbation is applied to represent all unstructured uncertainties in the system modeling. Moreover, the performance conditions in the damping ratio and the real part of the dominant mode is applied to formulate the optimization problem. In this work, the structure of the proposed controllers are the conventional first-order controller (lead/lag compensator). To achieve the controller parameters, the genetic algorithm (GA) is used to solve the optimization problem. Various simulation studies are carried out to confirm the performance of the proposed controller.

2. Proposed control design method

2.1 System uncertainties

System nonlinear characteristics, variations of system configuration due to unpredictable disturbances, loading conditions etc., cause various uncertainties in the power system. A controller which is designed without considering system uncertainties in the system modeling, the robustness of the controller against system uncertainties can not be guaranteed. As a result, the controller may fail to operate and lose stabilizing effect under various operating conditions. To enhance the robustness of power system damping controller against system uncertainties, the inverse additive perturbation (Gu et.al. 2005) is applied to represent all possible unstructured system uncertainties. The concept of enhancement of robust stability margin is used to formulate the optimization problem of controller parameters.

The feedback control system with inverse additive perturbation is shown in Fig.1. G is the nominal plant. K is the designed controller. For unstructured system uncertainties such as various generating and loading conditions, variation of system parameters and

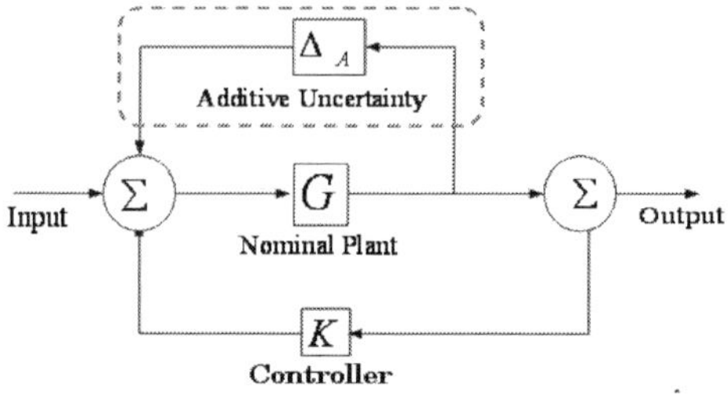

Fig. 1. Feedback system with inverse additive perturbation.

nonlinearities etc., they are represented by Δ_A which is the additive uncertainty model. Based on the small gain theorem, for a stable additive uncertainty Δ_A, the system is stable if

$$\left\|\Delta_A G / (1 - GK)\right\|_\infty < 1 \tag{1}$$

then,

$$\left\|\Delta_A\right\|_\infty < 1 / \left\|G / (1 - GK)\right\|_\infty \tag{2}$$

The right hand side of equation (2) implies the size of system uncertainties or the robust stability margin against system uncertainties. By minimizing $\left\|G/(1-GK)\right\|_\infty$, the robust stability margin of the closed-loop system is a maximum or near maximum.

2.2 Implementation

2.2.1 Objective function

To optimize the stabilizer parameters, an inverse additive perturbation based-objective function is considered. The objective function is formulated to minimize the infinite norm of $\left\|G/(1-GK)\right\|_\infty$. Therefore, the robust stability margin of the closed-loop system will increase to achieve near optimum and the robust stability of the power system will be improved. As a result, the objective function can be defined as

$$\text{Minimize} \qquad \left\|G/(1-GK)\right\|_\infty \tag{3}$$

It is clear that the objective function will identify the minimum value of $\left\|G/(1-GK)\right\|_\infty$ for nominal operating conditions considered in the design process.

2.2.2 Optimization problem

In this study, the problem constraints are the controller parameters bounds. In addition to enhance the robust stability, another objective is to increase the damping ratio and place the closed-loop eigenvalues of hybrid wind-diesel power system in a D-shape region (Abdel-Magid et.al. 1999). the conditions will place the system closed-loop eigenvalues in the D-shape region characterized by $\zeta \geq \zeta_{spec}$ and $\sigma \leq \sigma_{spec}$ as shown in Fig. 2.
Therefore, the design problem can be formulated as the following optimization problem.

$$\text{Minimize} \qquad \left\|G/(1-GK)\right\|_\infty \tag{4}$$

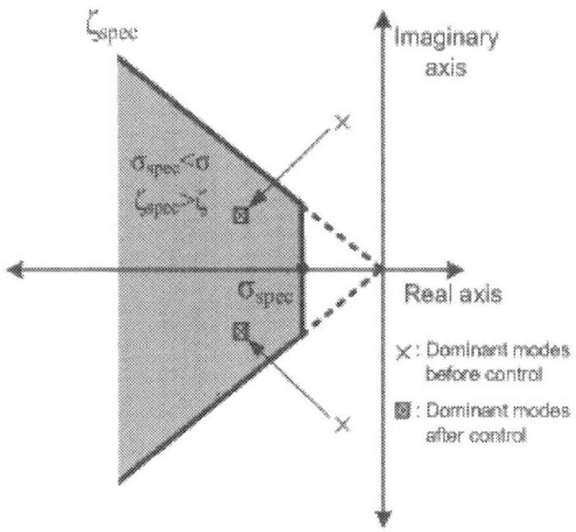

Fig. 2. D-shape region in the s-plane where $\sigma \le \sigma_{spec}$ and $\zeta \ge \zeta_{spec}$

Subject to
$$\zeta \ge \zeta_{spec}, \sigma \le \sigma_{spec} \tag{5}$$
$$K_{\min} \le K \le K_{\max}$$
$$T_{\min} \le T \le T_{\max}$$

where ζ and ζ_{spec} are the actual and desired damping ratio of the dominant mode, respectively; σ and σ_{spec} are the actual and desired real part, respectively; $K_{\max}$ and $K_{\min}$ are the maximum and minimum controller gains, respectively; $T_{\max}$ and $T_{\min}$ are the maximum and minimum time constants, respectively. This optimization problem is solved by GA (GAOT, 2005) to search the controller parameters.

2.3 Genetic algorithm

2.3.1 Overview

GA is a type of meta-heuristic search and optimization algorithms inspired by Darwin's principle of natural selection. GA is used to try and solving search problems or optimize existing solutions to a certain problem by using methods based on biological evolution. It has many applications in certain types of problems that yield better results than the common used methods.

According to Goldberg (Goldberg,1989), GA is different from other optimization and search procedures in four ways:

1. GA searches a population of points in parallel, not a single point.
2. GA does not require derivative information or other auxiliary knowledge; only the objective function and corresponding fitness levels influence the directions of search.
3. GA uses probabilistic transition rules, not deterministic ones.
4. GA works on an encoding of the parameter set rather than the parameter set itself (except in where real-valued individuals are used).

It is important to note that the GA provides a number of potential solutions to a given problem and the choice of final solution is left to the user.

2.3.2 GA algorithm

A. Representation of Individual.

Individual representation scheme determines how the problem is structured in the GA and also determines the genetic operators that are used. Each individual is made up of a sequence of genes. Various types of representations of an individual are binary digits, floating point numbers, integers, real values, matrices, etc. Generally, natural representations are more efficient and produce better solutions. Encoding is used to transform the real problem to binary coding problem which the GA can be applied.

B. GA Operators.

The basic search mechanism of the GA is provided by the genetic operators. There are two basic types of operators: crossover and mutation. These operators are used to produce new solutions based on existing solutions in the population. Crossover takes two individuals to be parents and produces two new individuals while mutation alters one individual to produce a single new solution (S. Panda,2009).
In crossover operator, individuals are paired for mating and by mixing their strings new individuals are created. This process is depicted in Fig. 3.

Parent 1	11010	01100
Parent 2	10110	11011
Child 1	11010	11011
Child 2	10110	01100

Fig. 3. Crossover operator

In natural evolution, mutation is a random process where one point of individual is replaced by another to produce a new individual structure. The effect of mutation on a binary string is illustrated in Fig. 4 for a 10-bit chromosome and a mutation point of 5 in the binary string. Here, binary mutation flips the value of the bit at the loci selected to be the mutation point (Andrew C et.al).

Parent 1	11010	01100
Child 1	11010	11100

Fig. 4. Mutation operator

C. Selection for Reproduction

To produce successive generations, selection of individuals plays a very significant role in a GA. The selection function determines which of the individuals will survive and move on to the next generation. A probabilistic selection is performed based upon the individual's fitness such that the superior individuals have more chance of being selected (S. Panda et.al ,2009). There are several schemes for the selection process: roulette wheel selection and its extensions, scaling techniques, tournament, normal geometric, elitist models and ranking methods. Roulette wheel selection method has simple method. The basic concept of this method is " High fitness, high chance to be selected".

2.3.3 Parameters optimization by GA

In this section, GA is applied to search the controller parameters with off line tuning. Each step of the proposed method is explained as follows.

Step 1. Generate the objective function for GA optimization.
In this study, the performance and robust stability conditions in inverse additive perturbation design approach is adopted to design a robust controller as mention in equation (4) and (5).

Step 2. Initialize the search parameters for GA. Define genetic parameters such as population size, crossover, mutation rate, and maximum generation.

Step 3. Randomly generate the initial solution.

Step 4. Evaluate objective function of each individual in equation (4) and (5).

Step 5. Select the best individual in the current generation. Check the maximum generation.

Step 6. Increase the generation.

Step 7. While the current generation is less than the maximum generation, create new population using genetic operators and go to step 4. If the current generation is the maximum generation, then stop.

3. Robust frequency control in a hybrid wind-diesel power system

3.1 System modeling

The basic system configuration of an isolated hybrid wind-diesel power generation system as shown in Fig. 5 (Das et.al. 1999) is used in this study. The base capacity of the system is 350 kVA. The diesel is used to supply power to system when wind power could not adequately provide power to customer. Moreover, The PPC is installed in the wind side while the governor is equipped with the diesel side. In addition to the random wind energy supply, it is assumed that loads with sudden change have been placed in this isolated system. These result in a serious problem of large frequency deviation in the system. As a result, a serious problem of large frequency deviation may occur in the isolated power system. Such power variations and frequency deviations severely affect the system stability. Furthermore, the life time of machine apparatuses on the load side affected by such large frequency deviations will be reduced.

3.2 Pitch control design in a hybrid wind-diesel power system

3.2.1 Linearized model of hybrid wind-diesel power system with PPC

For mathematical modelling, the transfer function block diagram of a hybrid wind-diesel power generation used in this study is shown in Fig. 6 (Das et.al. 1999). The PPC is a 1st order lead-lag controller with single input feedback of frequency deviation of wind side.
The state equation of linearized model in Fig. 6 can be expressed as

$$\Delta \dot{X} = A\Delta X + B\Delta u_{PPC} \tag{6}$$

$$\Delta Y = C\Delta X + D\Delta u_{PPC} \tag{7}$$

$$\Delta u_{PPC} = K(s)\Delta f_W \tag{8}$$

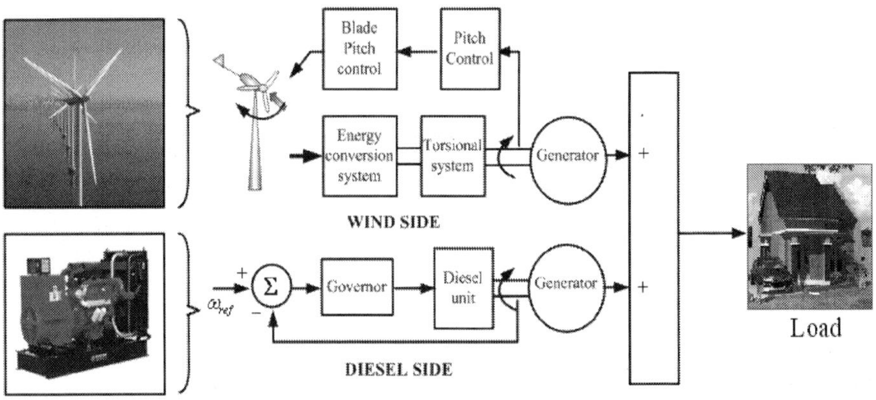

Fig. 5. Basic configuration of a hybrid wind-diesel power generation system.

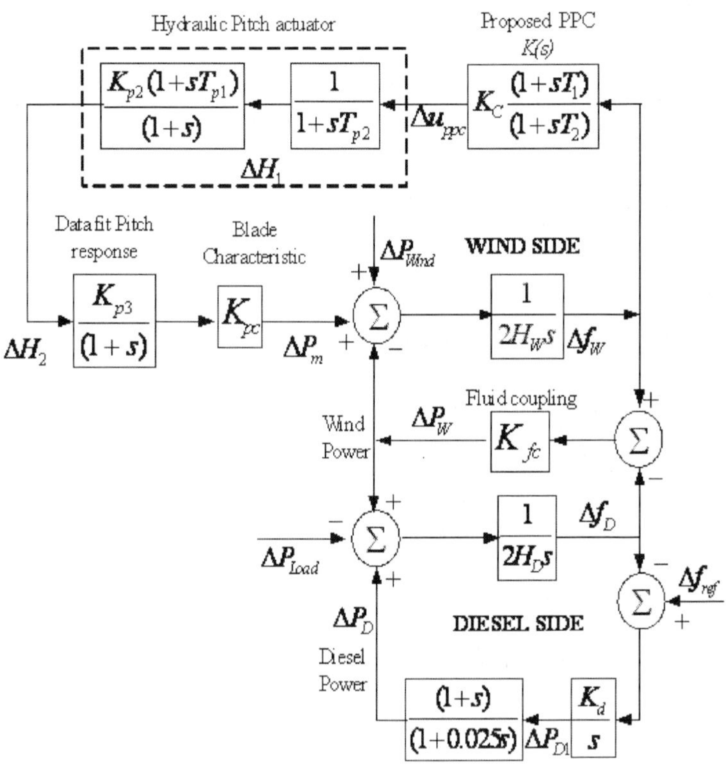

Fig. 6. Functional block diagram for wind-diesel system with proposed PPC.

Where the state vector $\Delta X = [\Delta f_W \quad \Delta f_D \quad \Delta P_{D1} \quad \Delta P_D \quad \Delta H_1 \quad \Delta H_2 \quad \Delta P_m]$, the output vector $\Delta Y = [\Delta f_W]$, ΔU_{PPC} is the control output of the PPC. The proposed control is applied to design a proposed PPC *K(s)*. The system in equation (6) is referred to as the nominal plant *G*.

3.2.2 Optimization problem formulation

The optimization problem can be formulated as follows,

Minimize $$\left\| G/(1-GK) \right\|_\infty \qquad (9)$$

Subject to $$\zeta \geq \zeta_{spec}, \sigma \leq \sigma_{spec} \qquad (10)$$

$$K_{\min} \leq K \leq K_{\max}$$

$$T_{\min} \leq T \leq T_{\max}$$

where ζ and ζ_{spec} are the actual and desired damping ratio of the dominant mode, respectively; σ and σ_{spec} are the actual and desired real part, respectively; $K_{\max}$ and $K_{\min}$ are the maximum and minimum controller gains, respectively; $T_{\max}$ and $T_{\min}$ are the maximum and minimum time constants, respectively. This optimization problem is solved by GA to search optimal or near optimal set of the controller parameters.

3.2.3 Designed results

In this section, simulation studies in a hybrid wind-diesel power generation are carried out. System parameters are given in (Das et.al. 1999). In the optimization, the ranges of search parameters and GA parameters are set as follows: $K_C \in [1 \quad 100]$, T_1 and $T_2 \in [0.0001 \quad 1]$, crossover probability is 0.9, mutation probability is 0.05, population size is 200 and maximum generation is 100. As a result, "the proposed PPC" is given automatically.

In simulation studies, the performance and robustness of the proposed PPC is compared with those of the PPC designed by the variable structure control (VSC) obtained from (Das et.al. 1999). Simulation results under four case studies are carried out as shown in table 1.

Cases	Disturbances
1	Step input of wind power or load change
2	Random wind power input
3	Random load power input
4	Simultaneous random wind power and load change.

Table 1. Operating conditions

Case 1: Step input of wind power or load change

First, a 0.01 pukW step increase in the wind power input and the load power are applied to the system at t = 5.0 s, respectively. Fig. 7 and Fig. 8 show the frequency deviation of the diesel generation side which represents the system frequency deviation. The peak frequency deviation is reduced significantly by both of the VSC PPC and the proposed PPC. However, the proposed PPC is able to damp the peak frequency deviation quickly in comparison to VSC PPC cases.

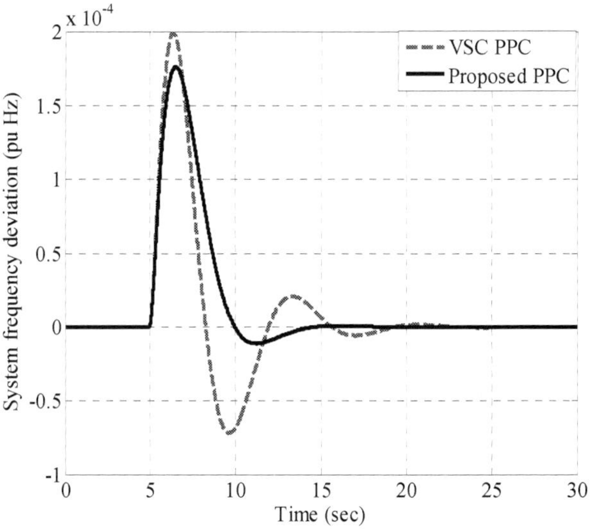

Fig. 7. System frequency deviation against a step change of wind power.

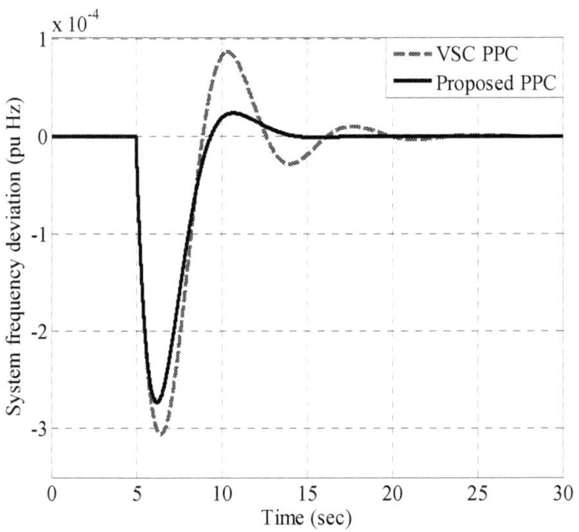

Fig. 8. System frequency deviation against a step load change.

Case 2: Random wind power input.

In this case, the system is subjected to the random wind power input as shown in Fig.9. The response of system frequency deviation is shown in Fig.10. By the proposed PPC, the frequency deviation is significantly reduced in comparison to that of the VSC PPC.

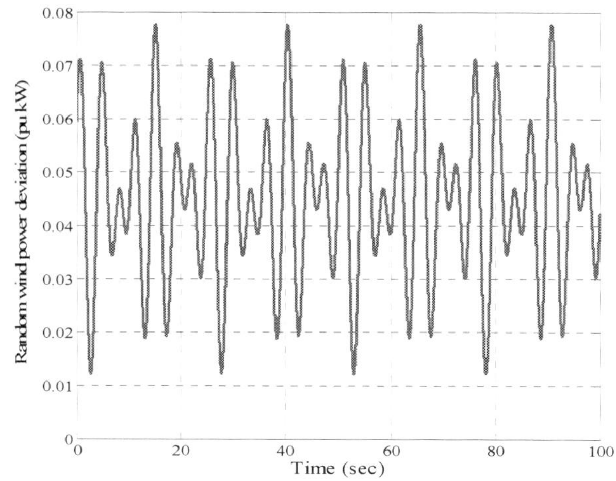

Fig. 9. Random wind power input.

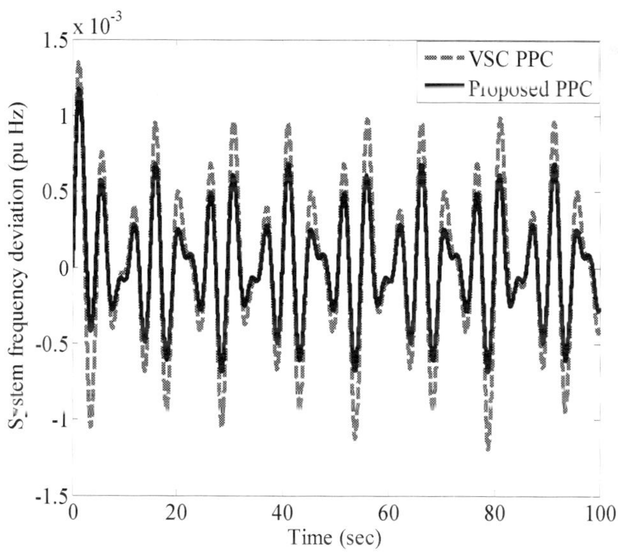

Fig. 10. System frequency deviation in case 2

Case 3: Random load change.

Next. the random load change as shown in Fig.11 is applied to the system. Fig. 12 depicts the response of system frequency deviation under the load change disturbance. The control effect of the proposed PPC is better than that of the VSC PPC.

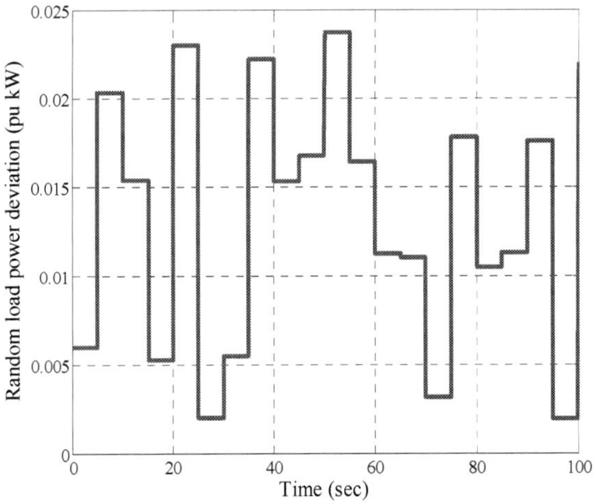

Fig. 11. Random load change

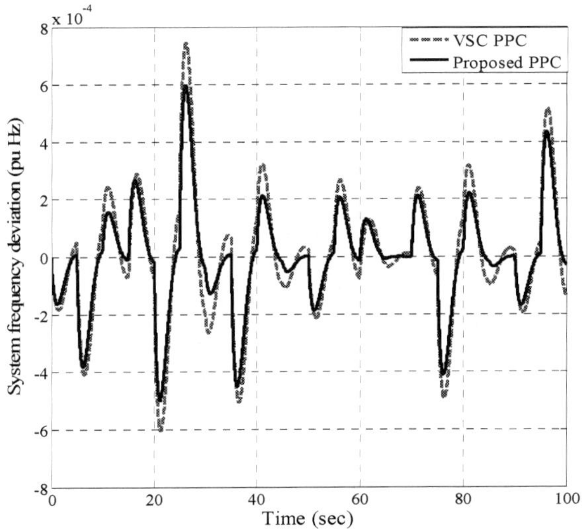

Fig. 12. System frequency deviation in case 3.

Case 4: Simultaneous random wind power and load change.

In this case, the random wind power input in Fig. 9 and the load change in Fig.11 are applied to the hybrid wind-diesel power system simultaneously. The response of system frequency deviation is shown in Fig. 13. The frequency control effect of the proposed PPC is superior to that of the VSC PPC.

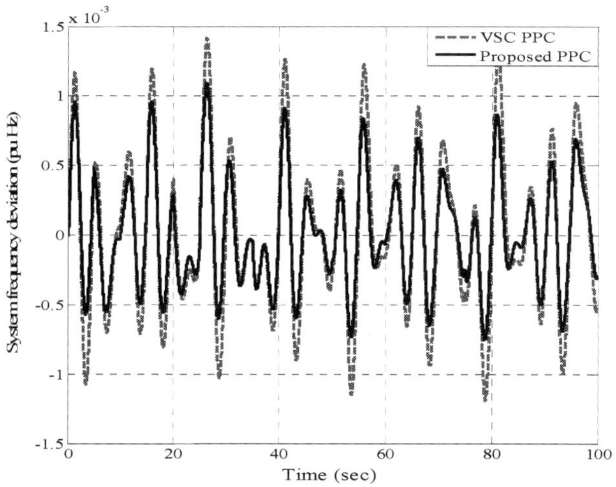

Fig. 13. System frequency deviation in case 4.

3.3 Frequency control in a hybrid wind-diesel power system using SMES

In this study, the system configuration in Fig. 5 is used to design frequency controller using SMES. In worst case, it is assumed that the ability of the pitch controller in the wind side and the governor in the diesel side to provide frequency control are is not adequate due to theirs slow response. Accordingly, the SMES is installed in the system to fast compensate for surplus or insufficient power demands, and minimize frequency deviation. Here, the proposed method is applied to design the robust frequency controller of SMES.

3.3.1 Linearized model of hybrid wind-diesel power system with PPC and SMES

The linearized model of the hybrid wind-diesel power system with Programmed Pitch Controller (PPC) and SMES is shown in Fig.14 (Tripathy, 1997). This model consists of the following subsystems: wind dynamic model, diesel dynamic model, SMES unit, blade pitch control of wind turbine and generator dynamic model. The details of all subsystems are explained in (Tripathy, 1997). As shown in Fig. 15, the SMES block diagram consists of two transfer functions, i.e. the SMES model and the frequency controller. Based on (Mitani et.al. 1988), the SMES can be modeled by the first-order transfer function with time constant $T_{sm} = 0.03$ s. In this work, the frequency controller is practically represented by a lead/lag compensator with first order. In the controller, there are three control parameters i.e., K_{sm}, T_{sm1} and T_{sm2}.

The linearized state equation of system in Fig. 14 can be expressed as

$$\Delta \dot{X} = A\Delta X + B\Delta u_{SM} \tag{11}$$

$$\Delta Y = C\Delta X + D\Delta u_{SM} \tag{12}$$

$$\Delta u_{SM} = K_{SM}\Delta u_{IN} \tag{13}$$

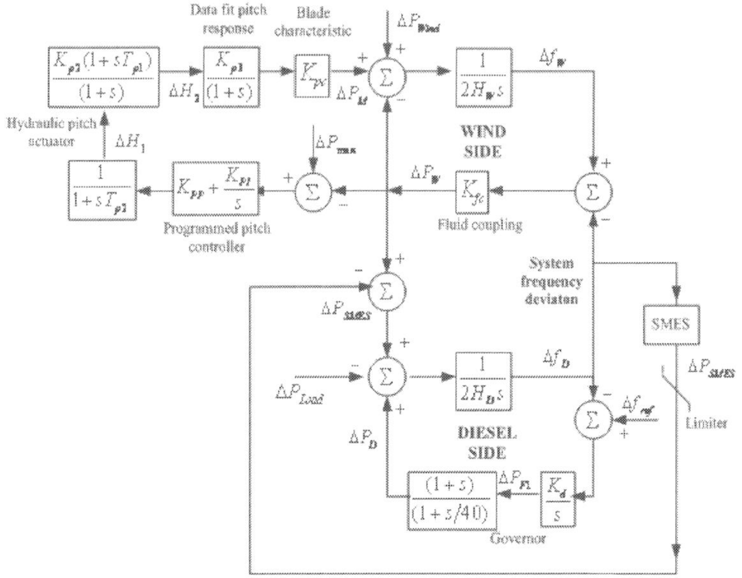

Fig. 14. Block diagram of a hybrid wind-diesel power generation with SMES.

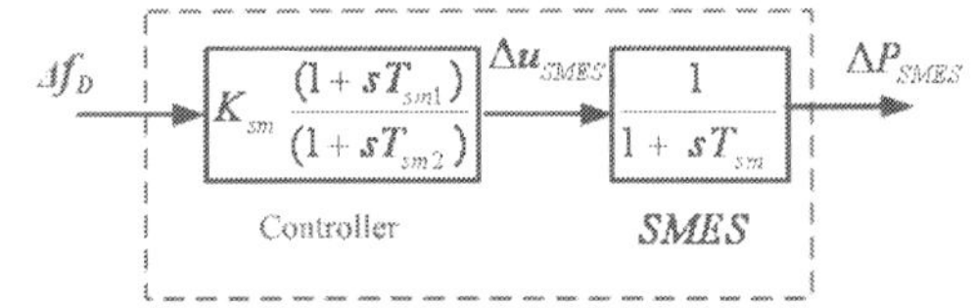

Fig. 15. Block diagram of SMES with the frequency controller.

Where the state vector $\Delta X = [\Delta f_W \ \ \Delta f_D \ \ \Delta P_{F1} \ \ \Delta P_D \ \ \Delta H_0 \ \ \Delta H_1 \ \ \Delta H_2 \ \ \Delta P_M]^T$, the output vector $\Delta Y = [\Delta f_D]$, Δf_D is the system frequency deviation, ΔP_{SMSES} is the control output signal of SMES controller; $\Delta u_{IN} = [\Delta Y]$ is the feedback input signal of SMES controller. Note that the system in equation (11) is a single-input single-output (SISO) system. The proposed method is applied to design SMES controller, and the system of equation (11) is referred to as the nominal plant G..

3.3.2 Optimization problem formulation

The optimization problem can be formulated as follows,

Minimize $$\|G/(1-GK)\|_\infty \quad (14)$$

Subject to $$\zeta \geq \zeta_{spec}, \sigma \leq \sigma_{spec} \quad (15)$$

$$K_{\min} \le K \le K_{\max}$$

$$T_{\min} \le T \le T_{\max}$$

where ζ and ζ_{spec} are the actual and desired damping ratio of the dominant mode, respectively; σ and σ_{spec} are the actual and desired real part, respectively; $K_{\max}$ and $K_{\min}$ are the maximum and minimum controller gains, respectively; $T_{\max}$ and $T_{\min}$ are the maximum and minimum time constants, respectively. This optimization problem is solved by GA to search optimal or near optimal set of the controller parameters.

3.3.3 Designed results

In the optimization, the ranges of search parameters and GA parameters are set as follows: ζ_{spec} is desired damping ratio is set as 0.4, σ_{spec} is desired real part of the dominant mode is set as -0.2, and $K_{\min}$ are $K_{\max}$ minimum and maximum gains of SMES are set as 1 and 60, $T_{\min}$ and $T_{\max}$ are minimum and maximum time constants of SMES are set as 0.01 and 5. The optimization problem is solved by genetic algorithm. As a result, the proposed controller which is referred as "RSMES" is given.

Table 2 shows the eigenvalue and damping ratio for normal operating condition. Clearly, the desired damping ratio and the desired real part are achieved by RSMES. Moreover, the damping ratio of RSMES is improved as designed in comparison with No SMES case.

Cases	Eigenvalues (damping ratio)
NO SMES	-39.0043 -24.4027 -3.5072 -1.2547 $-0.1851 \pm j\,0.671$, $\xi = 0.266$ $-0.5591 \pm j\,0.541$, $\xi = 0.719$
RSMES	-39.5266 -24.4006 -2.1681 -1.3325 $-17.782 \pm j\,5.339$, $\xi = 0.958$ $0.3050 \pm j\,0.539$, $\xi = 0.492$ $-0.2012 \pm j\,0.268$, $\xi = 0.600$

Table 2. Eigenvalues and Damping ratio

To evaluate performance of the proposed SMES, simulation studies are carried out under four operating conditions as shown in Table 1. In simulation studies, the limiter -0.01 pukW $\le \Delta P_{SMES} \le 0.01$ pukW on a system base 350 kVA is added to the output of SMES with each controller to determine capacity of SMES. The performance and robustness of the proposed controllers are compared with the conventional SMES controllers (CSMES) obtained from (Tripathy,1997). Simulation results under 4 case studies are carried out as follows.

Case 1: Step input of wind power or load change

In case 1, a 0.01 pukW step increase in the wind power input are applied to the system at t = 0.0 s. Fig. 16 shows the frequency deviation of the diesel generation side which represents the system frequency deviation. Without SMES, the peak frequency deviation is very large. The frequency deviation takes about 25 s to reach steady-state. This indicates that the pitch controller in the wind side and the governor in the diesel side do not work well. On the other hand, the peak frequency deviation is reduced significantly and returns to zero within shorter period in case of CSMES and the RSMES. Nevertheless, the overshoot and setting time of frequency oscillations in cases of RSMES is lower than that of CSMES.

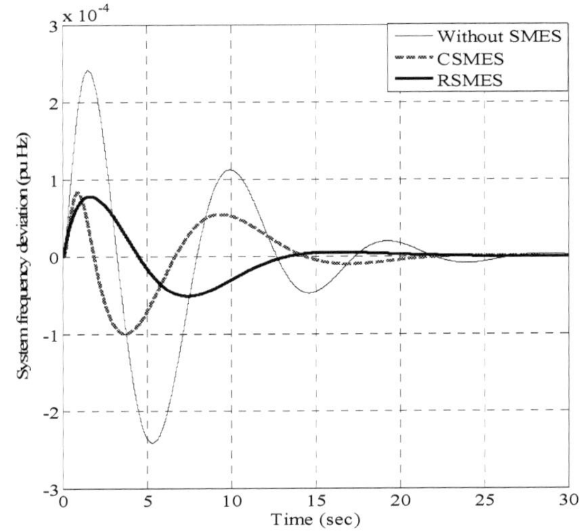

Fig. 16. System frequency deviation against a step change of wind power.

Next, a 0.01 pukW step increase in the load power is applied to the system at t = 0.0 s. As depicted in Fig. 17, both CSMES and RSMES are able to damp the frequency deviation quickly in comparison to without SMES case. These results show that both CSMES and RSMES have almost the same frequency control effects.

Case 2: Random wind power input.

In this case, the system is subjected to the random wind power input as shown in Fig.18. The system frequency deviations under normal system parameters are shown in Fig.19. Normal system parameter is the design point of both CSMES and RSMES. By the RSMES, the frequency deviation is significantly reduced in comparison to that of CSMES.
Next, the robustness of frequency controller is evaluated by an integral square error (ISE) under variations of system parameters. For 100 seconds of simulation study under the same random wind power in Fig.18, the ISE of the system frequency deviation is defined as

$$\text{ISE of} \quad \Delta f_D = \int_0^{100} |\Delta f_D|^2 \, dt \qquad (16)$$

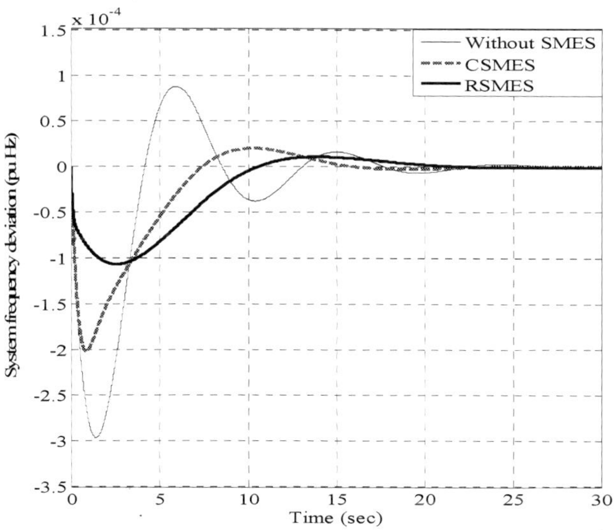

Fig. 17. System frequency deviation against a step load change.

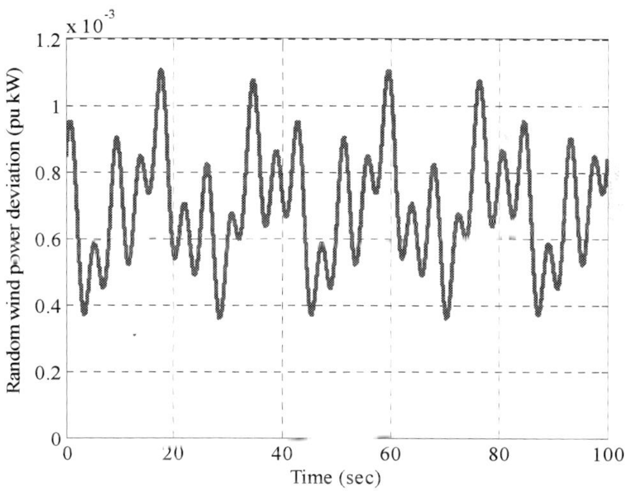

Fig. 18. Random wind power input.

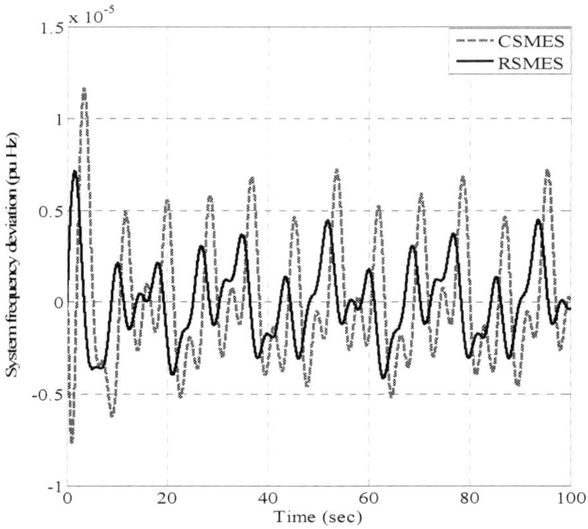

Fig. 19. System frequency deviation under normal system parameters.

Fig.20 shows the values of ISE when the fluid coupling coefficient K_{fc} is varied from -30 % to +30 % of the normal values. The values of ISE in case of CSMES largely increase as K_{fc} decreases. In contrast, the values of ISE in case of RSMES are lower and slightly change.

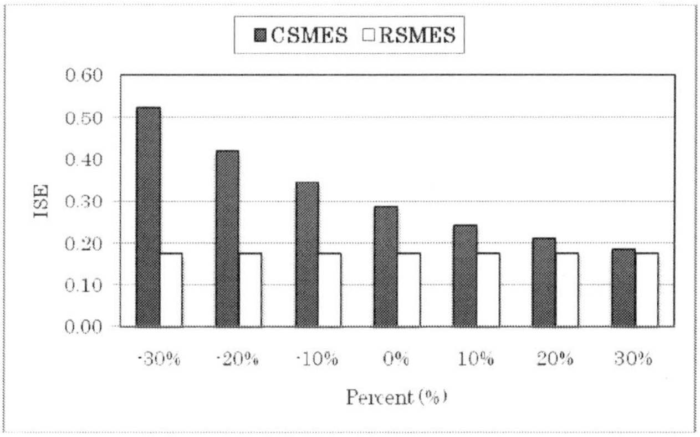

Fig. 20. Variation of ISE under a change of K_{fc}.

Case 3: Random load change.

Fig. 22 shows the system frequency deviation under normal system parameters when the random load change as shown in Fig.21 is applied to the system. The control effect of RSMES is better than that of the CSMES.

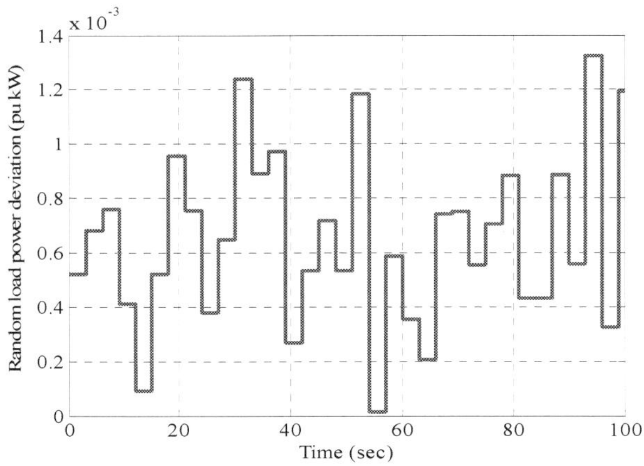

Fig. 21. Random load change.

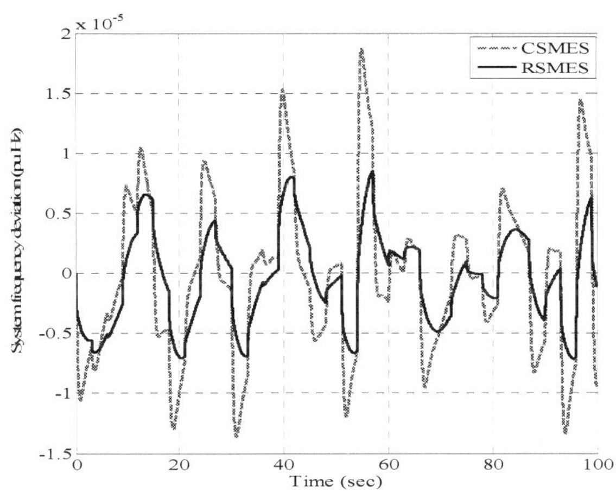

Fig. 22. System frequency deviation under normal system parameters.

Case 4: Simultaneous random wind power and load change.

In case 4, the random wind power input in Fig. 18 and the load change in Fig.21 are applied to the system simultaneously. When the inertia constant of both sides are reduced by 30 % from the normal values, the CSMES is sensitive to this parameter change. It is still not able to work well as depicted in Fig.23. In contrast, RSMES is capable of damping the frequency oscillation. The values of ISE of system frequency under the variation of K_{fc} from -30 % to +30 % of the normal values are shown in Fig.24. As K_{fc} decreases, the values of ISE in case of CSMES highly increase. On the other hand, the values of ISE in case of RSMES are much lower and almost constant. These simulation results confirm the high robustness of RSMES against the random wind power, load change, and system parameter variations.

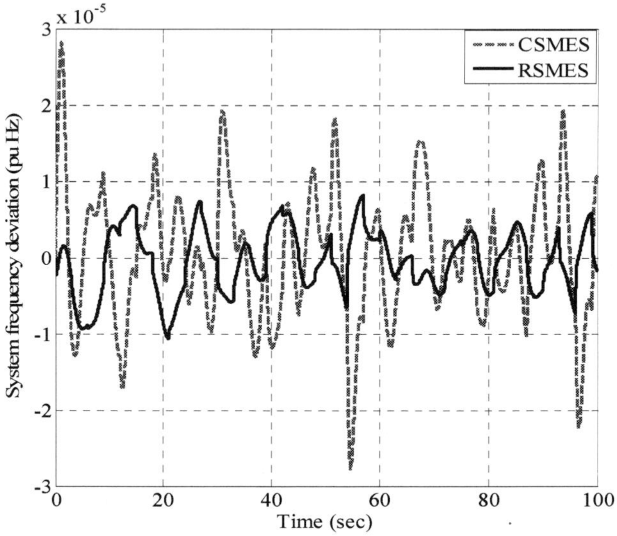

Fig. 23. System frequency deviation under a 30 % decrease in K_{fc}

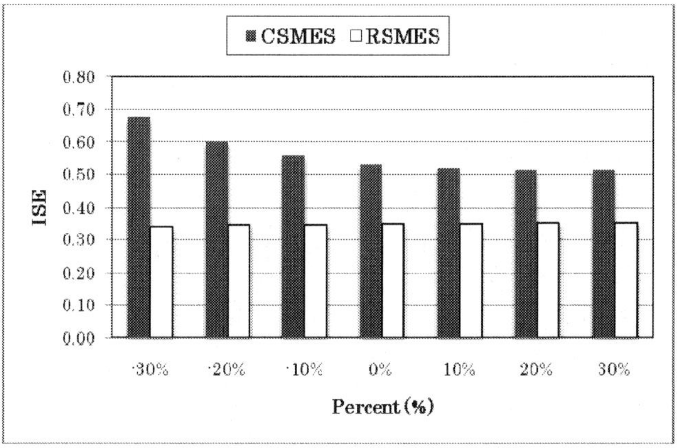

Fig. 24. Variation of ISE under a change in K_{fc}.

Finally, SMES capacities required for frequency control are evaluated based on simultaneous random wind power input and load change in case study 4 in addition to a 30 % decrease in K_{fc} parameters. The kW capacity is determined by the output limiter $-0.01 \leq \Delta P_{SMES} \leq 0.01$ pukW on a system base of 350 kW. The simulation results of SMES output power in case study 4 are shown in Figs. 25. Both power output of CSMES and RSMES are in the allowable limits. However, the performance and robustness of frequency oscillations in cases of RSMES is much better than those of CSMES.

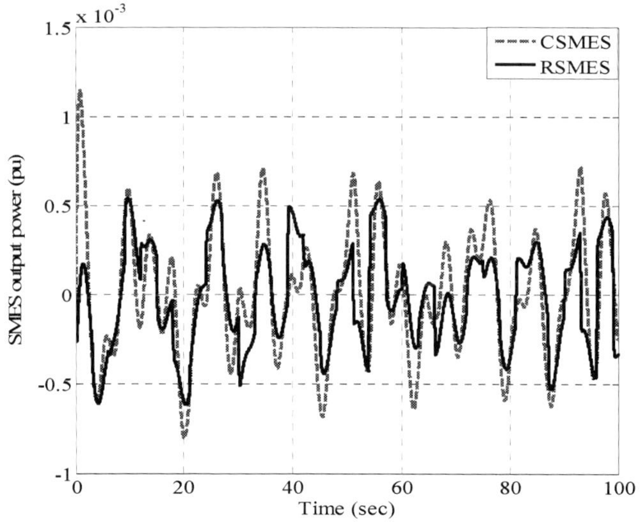

Fig. 25. SMES output power under a 30 % decrease in K_{fc}

5. Conclusion

Control scheme of hybrid wind-diesel power generation has been proposed in this work. This work focus on frequency control using robust controllers such as Pitch controller and SMES. The robust controllers were designed based on inverse additive perturbation in an isolated hybrid wind - diesel power system. The performance and stability conditions of inverse additive perturbation technique have been applied as the objective function in the optimization problem. The GA has been used to tune the control parameters of controllers. The designed controllers are based on the conventional 1st-order lead-lag compensator. Accordingly, it is easy to implement in real systems. The damping effects and robustness of the proposed controllers have been evaluated in the isolated hybrid wind - diesel power system. Simulation results confirm that the robustness of the proposed controllers are much superior to that of the conventional controllers against various uncertainties.

6. References

Ackermann, T. (2005), Wind Power in Power Systems, *John Wiley & Sons.*

Hunter R.E.G. (1994), Wind-diesel systems a guide to technology and its implementation, *Cambridge University Press.*

Lipman NH. (1989), Wind-diesel and autonomous energy systems, *Elservier Science Publishers Ltd.*

Bhatti T.S., Al-Ademi A.A.F. & Bansal N.K. (1997), Load frequency control of isolated wind diesel hybrid power systems, *International Journal of Energy Conversion and Management*, Vol. 39, pp. 829-837.

Das, D., Aditya, S.K., & Kothari, D.P. (1999), Dynamics of diesel and wind turbine generators on an isolated power system, *International Journal of Elect Power & Energy Syst.*, vol. 21, pp.183-189.

Mohan Mathur R & Rajiv K. Varma. (2002), Thyristor-based FACTS controllers for electrical transmission Systems, *John Wiley*.

Ribeiro P.F. , Johnson B.K., Crow M.L., Arsoy A. & Liu Y. (2001), Energy Storage Systems for Advanced Power Applications, *Proc. of the IEEE*, Vol. 89, No. 12, pp. 1744 -1756.

Schainker R.B. (2004), Executive Overview: Energy Storage Options for a Sustainable Energy Future, *IEEE Power Engineering Society General Meeting*, pp. 2309 - 2314.

Jiang X. & Chu X. (2001), SMES System for Study on Utility and Customer Power Applications, *IEEE Trans. on Applied Superconductivity*, Vol. 11, pp. 1765–1768.

Simo J.B. & Kamwa I. (1995), Exploratory Assessment of the Dynamic Behavior of Multi-machine System Stabilized by a SMES Unit, *IEEE Trans. on Power Systems*, Vol.10, No.3, pp. 1566–1571.

Wu C.J. &. Lee Y.S. (1993), Application of Simultaneous Active and Reactive Power Modulation of Superconducting Magnetic Energy Storage Unit to Damp Turbine-Generator Subsynchronous Oscillations, *IEEE Trans. on Energy Conversion*, Vol.8, No.1, pp. 63–70.

Maschowski J. & Nelles D. (1992), Power System Transient Stability Enhancement by Optimal Simultaneous Control of Active and Reactive Power, *IFAC symposium on power system and power plant control*, Munich, pp. 271–276.

Buckles W. & Hassenzahl W.V. (2000), Superconducting Magnetic Energy Storage, *IEEE Power Engineering Review*, pp.16–20.

Juengst K.P. (1998), SMES Progress, *Proc. of 15th International Conference on Magnet Technology, Science Press*, pp. 18–23.

Rabbani M.G., Devotta J.B.X. & Elangovan . (1998), An Adaptive Fuzzy Controlled Superconducting Magnetic Energy Storage Unit for Power Systems, *Energy Conversion and Management* , Vol. 39, pp.931-942.

Devotta J.B.X. & Rabbani M.G. (2000), Application of Superconducting Magnetic Energy Storage Unit in Multi-machine Power Systems, *Energy Conversion and Management*, Vol. 41, pp. 493-504.

Tripathy S.C. (1997), Dynamic Simulation of Hybrid Wind-diesel Power Generation System with Superconducting Magnetic Energy Storage, *Energy Conversion and Management*, Vol.38 , pp.919-930.

Ngamroo . (2005), An Optimization Technique of Robust Load Frequency Stabilizers for Superconducting Magnetic Energy Storage, *Energy Conversion and Management*, Vol.46, pp.3060-3090.

Chu X., Jiang X., Lai Y., Wu X. & Liu W. (2001), SMES Control Algorithms for Improving Customer Power Quality, *IEEE Trans. on Applied Superconductivity*, Vol. 11, pp.1769-1772.

Devotta J.B.X., Rabbani M.G. & Elangovan S. (1999), Application of Superconducting Magnetic Energy Storage Unit for Damping of Subsynchronous Oscillations in Power Systems, *Energy Conversion and Management*, Vol.40, pp.23-37.

Abdelsalam M.K., Boom R.W & Perterson H.A. (1987) , Operation Aspects of Superconducting Magnetic Energy Storage (SMES), *IEEE Trans. on Magnetics*, Vol.23, pp. 3275-3277.

Banerjee S., Chatterjee J.K & Tripathy S.C. (1990), Application of Magnetic Energy Storage Unit as Load Frequency Stabilizer, *IEEE Trans. on Energy Conversions*, Vol. 5, No. 1, pp. 46–51.

Tripathy S.C. & Juengst K.P. (1997), Sampled Data Automatic Generation Control with Superconducting Magnetic Energy Storage in Power Systems, *IEEE Trans. on Energy Conversions*, Vol. 12, No. 2, pp. 187–191.

Tripathy S.C., Balasubramanian R. & Chandramohanan Nair P.S. (1997), Adaptive Automatic Generation Control with Superconducting Magnetic Energy Storage in Power Systems, *IEEE Trans. on Energy Conversion*, Vol. 7, No. 3, pp. 434–441.

Demiroren A., Zeynelgil H.L & Sengor N.S. (2003), Automatic Generation Control for Power System with SMES by using Neural Network Controller, *Electrical Power Components Systems*, Vol. 31, No.1, pp. 1–25.

Demiroren A. & Yesil E. (2004), Automatic Generation Control with Fuzzy Logic Controllers in the Power System Including SMES Units, *Electrical Power Energy Systems*, Vol. 26, pp. 291–305.

Djukanovic M., Khammash M. & Vittal V. (1999), Sequential Synthesis of Structured Singular Values Based Decentralized Controllers in Power Systems, *IEEE Trans. on Power Systems*, Vol. 14, No. 2, pp. 635–641.

Yu X., Khammash M. & Vittal V. (2001), Robust Design of a Damping Controller for Static Var Compensators in Power Systems, *IEEE Trans. on Power Systems*, Vol. 16, No.3, pp.456–462.

Zhu C., Khammash M. , Vittal V. & Qui W. (2003) , Robust Power System Stabilizer Design using H_∞ Loop Shaping Approach, *IEEE Trans. on Power Systems*, Vol. 18, No.2, pp. 810–818.

Rahim A.H M.A. & Kandlawala M.F. (2004), Robust STATCOM Voltage Controller Design using Loop-shaping Technique, *Trans. on Electric Power Systems Research*, Vol. 68 No.1, pp. 61–74.

Wang Y., Tan Y.L. & Guo G. (2002), Robust Nonlinear Coordinated Excitation and TCSC Control for Power Systems, *IEE Proc. of Generation Transmission and Distribution*, Vol. 149, No. 3, pp. 367–372.

Tan Y.L. & Wang Y. (2004), A Robust Nonlinear Excitation and SMES Controller for Transient Stabilization, *Electrical Power Energy Systems*, Vol. 26, No.5, pp. 325–332.

Gu P., Petkov Hr. & Konstantinov M.M.(2005), Robust Control Design with MATLAB, *Springer*.

Abdel-Magid Y. L., Abido M. A., AI-Baiyat S. & Mantawy A. H. (1999), Simultaneous Stabilization of Multimachine Power Systems via Genetic Algorithm, *IEEE Trans. on Power Systems*, Vol. 14, No. 4, pp. 1428-1439.

GAOT (2005), A Genetic Algorithm for Function Optimization: A Matlab Implementation. [Online] Available: http://www.ie.ncsu.edu/mirage/GAToolBox/gaot/

Goldberg D.E. (1989), Genetic Algorithm in Search, Optimization and Machine Learning, *Addison-Wesley Publishing Company Inc.*

Das D., Aditya, S.K., & Kothari, D.P. (1999), Dynamics of diesel and wind turbine generators on an isolated power system, *International Journal of Elect Power & Energy Syst.*, vol. 21, pp.183-189.

Tripathy SC . (1997), Dynamic simulation of hybrid wind-diesel power generation system with superconducting magnetic energy storage, *Energy Conv and Manag.* Vol. 38, No. 9, 919-930.

Tripathy SC, Kalantar M & Balasubramanian R. (1991), Dynamic and stability of wind and diesel turbine generators with superconducting magnetic energy storage unit on an isolated power system, *IEEE Trans on Energy Conv* , Vol. 6, No. 4, pp. 579-585.

Mitani Y, Tsuji K & Murakami Y. (1988), Application of superconducting magnetic energy storage to improve power system dynamic performance , *IEEE Trans. Power Syst* , Vol. 3, No. 4, pp. 1418-25.

Panda S, Yadav J.S, Patidar N.P and Ardil C. (2009), Evolutionary Techniques for Model Order Reduction of Large Scale Linear Systems, *International Journal of Engineering and Applied Sciences*, Vol. 5, No.1, pp. 22-28.

Andrew C, Peter F, Hartmut P and Carlos F. Genetic Algorithm TOOLBOX For Use with MATLAB- User's guide, Department of automatic control and systems engineering, university of Sheffield.

Power Fluctuations in a Wind Farm Compared to a Single Turbine

Joaquin Mur-Amada and Jesús Sallán-Arasanz
Zaragoza University
Spain

1. Introduction

This chapter is focused on the estimation of wind farm power fluctuations from the behaviour of a single turbine during continuous operation (special events such as turbine tripping, grid transients, sudden voltages changes, etc. are not considered). The time scope ranges from seconds to some minutes and the geographic scope is bounded to one or a few nearby wind farms.

One of the objectives of this chapter is to explain quantitatively the wind power variability in a farm from the behaviour of a single turbine. For short intervals and inside a wind farm, the model is based on the experience with a logger system designed and installed in four wind farms (Sanz et al., 2000a), the classic theory of Gaussian (normal) stochastic processes, the wind coherence model (Schlez & Infield, 1998), and the general coherence function derived by Risø Institute in Horns Rev wind farm (Martins et al., 2006; Sørensen et al., 2008a). For larger distances and slower variations, the model has been tested with meteorological data from the weather network.

The complexities inherent to stochastic processes are partially circumvented presenting some case studies with meaningful graphs and using classical tools of signal processing and time series analysis when possible. The sum of the power from many turbines is a stochastic process that is the outcome of many interactions from different sources. The sum of the power variations from more than four turbines converges approximately to a Gaussian process despite of the process nature (deterministic, stochastic, broadband or narrowband), analogously to the martingale central limit theorem (Hall & Heyde, 1980). The only required condition is the negligible effect of synchronization forces among turbine oscillations.

The data logged at some wind farms are smooth and they have good mathematical properties except during special events such as turbine breaker trips or severe weather. This chapter will show that, under some circumstances, the power output of a wind farm can be approximated to a Gaussian process and its auto spectrum density can be estimated from the spectrum of a turbine, wind farm dimensions and wind coherence. The wind farm power variability is fully characterized by its auto spectrum provided the Gaussian approximation is accurate enough. Many interesting properties such as the mean power fluctuation shape during a period, the distribution of power variation in a time period, the more extreme power variation expected during a short period, etc. can be estimated applying the outstanding properties of Gaussian processes according to (Bierbooms, 2008) and (Mur-Amada, 2009).

Since the canonical representation of a Gaussian stochastic process is its frequency spectrum (Karhunen–Loeve theorem), the analysis of wind power fluctuations is usually done in the frequency domain for convenience. An alternative to Fourier analysis is time series analysis. Time series are quite popular in stochastic models since they are well suited to prediction and their parameters and their properties can be easily estimated (Wangdee & Billinton, 2006). Even though the two mathematical techniques are quite related, the study of periodic behaviour is more direct through Fourier approach whereas the time series approach is more appropriate for the study of non-systematic behaviour.

1.1 Sources of wind power fluctuation

The fluctuations observed at the output of a turbine are the outcome of the interaction of wind turbulence with the complex turbine dynamics. For very slow fluctuations (corresponding to lower frequencies in the spectrum), the turbine regulation achieves its target and the turbine dynamics are negligible. Faster fluctuations (corresponding to higher frequencies) interact with the structural and drive-train vibrations. The complexity of the mechanical vibrations, the turbine control and the non-linearity of the generator power electronics interactions affects notably the generator electromagnetic torque and the turbine power fluctuations, especially in the frequency range from tenths of Hertzs to grid frequency.

There are many dynamic turbine models described in the literature. Most megawatt turbines share the following behaviour, considering the aerodynamic torque as the system input and the power injected in the grid as the system output (Soens, 2005; Comech-Moreno, 2007; Bianchi et al, 2006):

- Between cut-in and rated wind speeds, the turbine power usually behaves (with respect to the wind measured with an anemometer) as a low frequency first-order filter with a time constant between 1 and 10 s.
- Between rated and cut-out wind speeds, the turbine power usually behaves (with respect to the measured wind) as an asymmetric band pass filter of characteristic frequency around 0,3 Hz due to the combined effect of the slow action of the pitch/active stall and the quicker speed controllers.
- At some characteristic frequencies, the turbine mechanical vibrations, the power electronics and the generator dynamics modify the general trend of the power output spectrum with respect to the wind input.

There are many specific characteristics that impact notably the power fluctuations and their realistic reproduction requires a comprehensive model of each turbine. The details of the control, the structural details and the power electronics implemented in the turbines are proprietary and they are not publicity available. In contrast, the electrical power injected by a turbine can be measured easily.

Moreover, some fluctuations in power are not proportional to the fluctuations in wind or aerodynamic torque. Thus, the ratio of the output signal divided by the input signal in the frequency domain is not constant. However, a statistical linear model in the frequency can be used (Welfonder et al., 1997) although the system output is neither proportional to the input nor deterministic.

The approach taken in this chapter is primarily phenomenological: the power fluctuations during the continuous operation of the turbines are measured and characterized for timescales in the range of minutes to fractions of seconds. Thus, one contribution of this chapter is the experimental characterization of the power fluctuations of three commercial turbines. Some experimental measurements in the joint time-frequency domain are presented to test the mathematical model of the fluctuations and the variability of PSD is studied through spectrograms.

Other contribution of this chapter is the admittance of the wind farm: the oscillations from a wind farm are compared to the fluctuations from a single turbine, representative of the operation of the turbines in the farm. The partial cancellation of power fluctuations in a wind farm is estimated from the ratio of the farm fluctuation relative to the fluctuation of one representative turbine. Some stochastic models are derived in the frequency domain to link the overall behaviour of a large number of wind turbines from the operation of a single turbine.

This chapter is based mostly on the experience obtained designing, programming, assembling and analyzing two multipurpose measuring system installed in several wind farms (Sanz et at., 2000a; Mur-Amada, 2009). This measuring system has been the first prototype of a multipurpose data logger, now called AIRE (Analizador Integral de Recursos Energéticos), that is currently commercialized by Inycom and CIRCE Foundation.

1.2 Random and almost cyclic fluctuations

Power output fluctuations can be divided into almost cyclic components (tower shadow, wind shear, modal vibrations, etc.), wind farm weather dynamics (turbulence, boundary layer atmospheric stability, micrometeorological dynamics, etc.) and events (connection or disconnection of the turbine, change in generator configuration, etc.). The customary treatment of these fluctuations is done through Fourier transform.

Cyclic fluctuations due to tower shadow, wind shear, etc. present more systematic behaviour than weather related variations. Almost cyclic fluctuations are approximately periodic and they present quite definite frequencies. In this context, almost periodic means that the signal can be decomposed into a set of sinusoidal components with slow varying amplitudes (some of them non-harmonically related) and stationary noise (i.e., polycyclostationary signals). The frequencies in the signal vary slightly since the fluctuation amplitudes are not constant and the signal is not periodic in the conventional sense.

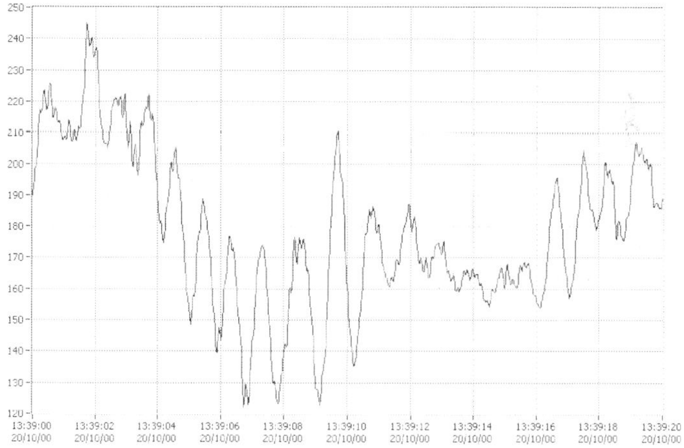

Fig. 1. Active power of a 750 kW wind turbine for wind speeds around 6,7 m/s during 20 s.

Cyclic variations are usually characterized with their Fourier transforms (Gardner, 1994). Moreover, turbulence is also characterized through its auto spectral density, which is basically the Fourier transform of its autocorrelation. Periodic fluctuations appear as narrow peaks at their harmonic frequencies in the spectrum, whereas random fluctuations (which have neither a periodic pattern nor a characteristic frequency) can be associated with the tendency of the smoothed spectrum. Thus, the magnitude and frequency of the cyclic fluctuations can be characterized for each turbine model and wind regime (Mur-Amada, 2009).

Weather evolution is the outcome of slow and complex atmospheric processes. Since weather evolution has a strong non-linear behaviour, it will not be considered in this chapter.

1.3 Fluctuations induced by the wind turbulence

Many fluctuations in the power output are strongly related to wind fluctuations, especially at low frequencies (slow fluctuations). The wind spectrum is a common way to characterize the frequency content of the turbulence present in the wind as it flows around an anemometer. The wind is usually measured in a fixed point, but the wind varies along a wind farm, not only due to the obstacles and orography, but also due to the turbulent nature of wind.

Taylor's hypothesis of frozen turbulence is a simple model that relates spatial and temporal variations of the wind. This hypothesis can be used to reconstruct the approximate spatial structure of wind from measurements with an anemometer fixed at a point in space.

In fact, wind irregularities experienced by a turbine are also perceived by the next turbines (usually with diverse magnitude and with some time delay). The area of influence of the turbulence is related to the value of wind speed deviations (Cushman-Roisin, 2007). Higher wind fluctuations usually imply larger spatial extent. Therefore, wind fluctuations are usually experienced in close turbines with some time lag/lead $\Delta t'$ In Taylor's Hypothesis of "frozen turbulence", the gust travel time in the wind direction $\Delta t'$ is the distance in longitudinal direction divided by the wind speed (see Fig. 2). The wind measured at the tower of Fig. 2 varies in 10 s due to a perturbation 100 m long travelling at the wind speed.

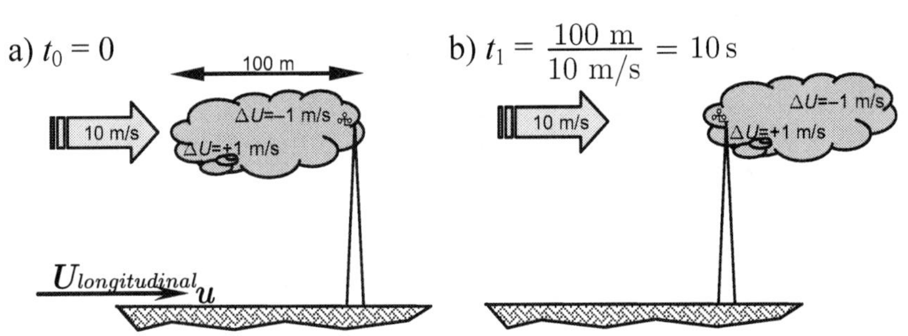

Fig. 2. Example of a idealized eddy of 100 m (represented by a cloud) passing through a meteorological mast according to Taylor's Hypothesis of "frozen turbulence".

If the fluctuation arrives to another turbine inside the time interval $[-\Delta t, +\Delta t]$, then the phase uncertainty in the frequency domain is $[-2\pi f \Delta t, +2\pi f \Delta t]$ radians, where f is the considered frequency. When $f > 0{,}5/\Delta t$, the phase is undetermined because the uncertainty of the phase excess $[-\pi, +\pi]$ (i.e. a cycle). At frequencies a few times higher than $0{,}5/\Delta t$, the fluctuation of frequency f is experienced by other turbines with a random phase difference almost uniformly distributed and with comparable amplitude. In other words, the phase of the fluctuations in the frequency domain are uncorrelated stochastically at $f > 0{,}5/\Delta t$ although the amplitude could show a systematic behaviour. The spatial and temporal coherence statistically quantifies the variations of wind in different points in space or in separate moments of time.

For convenience, the wind is sometimes assumed barely uniform in the area swept by the turbine. Based on this approximation, the *equivalent wind* is defined as the one that produces the same effects that the non-uniform real wind field. Although the wind field cannot be directly measured, its effects can be deduced from an equivalent wind that is usually derived from the measurements of an anemometer, because variations in time and space are related by the air flow dynamics.

The equivalent wind speed contains a stochastic component due to the effects of turbulence, a rotational component due to the wind shear and the tower shadow and the average value of the wind in the swept area, considered constant in short intervals. The rotational effects (wind shear and tower effect) are barely related to wind turbulence. Since they interact with the drive-train and control dynamics, they are modelled as an additional term in the oscillations. The rotational/vibration/control dynamics are introduced in the equivalent wind as a mathematical artifice to reproduce the power oscillations observed in the turbine output. This simplification works relatively well since the vibration turbine dynamics randomize the real dependence of the generator torque with the rotor angle.

The turbulence does not show characteristic frequencies and the wind spectrum is quite smooth from very low frequencies up to tenths of Hertzs. In contrast, rotational/vibration/control oscillations in the power output exhibit a more repetitive pattern with determinate characteristic frequencies. Apart from their frequency distribution, turbulence and other oscillations have similar stochastic properties and they can be modelled with the same mathematical tools.

The combination of the small signal model and the wind coherence permits to derive the spatial averaging of random wind variations. The stochastic behaviour of wind links the overall behaviour of a large number of turbines with the behaviour of a single turbine.

It should be noted that the travel time of the turbulence between the turbines is the very reason why fast fluctuations of turbine power generated by the turbulence are smoothed in the wind farm output. That is also the reason why a Gaussian processes is well suited to model the power fluctuations across a wind farm. Thus, the analysis carried out in this chapter is in the frequency domain for convenience. Moreover, this behaviour also relates the dimensions and geometry of the wind farm with the cut off frequency of the smoothing (the smoothing depends also on the wind coherence and direction).

The auto spectral density of the equivalent wind of a cluster of turbines can be obtained from the wind spectra, the parameters of an isolated turbine, lateral and longitudinal dimensions of the cluster region and the decay factor of the spatial coherence.

Fluctuations due to the real wind field along the swept area, vibrations and control effects are added to the equivalent wind modifying its spectra. Thus, they can be aggregated in the equivalent wind, provided a turbine transfer function among the power output and the equivalent wind is stated. The turbine transfer function transforms the equivalent wind oscillations into power oscillations. This simplification works relatively well since the turbine vibration dynamics randomize the turbine output and the high frequency turbulence at different turbines has a similar a stochastic behaviour than the rotational/vibration/control oscillations: at high frequencies, fluctuations from turbulence, vibration, generator dynamics and control are fairly independent between turbines, statistically speaking.

The combination of the small signal model and the wind coherence permits to derive the spatial averaging of random wind variations. Since fast turbulence and rotation/vibration/control oscillations are almost stochastically independent among the farm turbines, their outcome can be assessed analogously, although their respective sources are very different physical phenomena.

Thus, the overall behaviour of a turbine cluster (with more than 8 turbines) can be derived from the behaviour of a single turbine using a Gaussian model. The wind farm admittance is the ratio of the fluctuations observed in the farm output respect the typical behaviour of one of its turbines. The wind farm admittance can be estimated from experimental measurements or from parameters of an isolated turbine, lateral and longitudinal dimensions of the cluster region and the decay factor of the spatial coherence. Although the model proposed is an oversimplification of the actual behaviour of a group of turbines scattered across an area, this model quantifies the influence of the spatial distribution of the turbines in the smoothing and in the frequency content of the aggregated power. This stochastic model is in agreement with the experimental data presented at the end of this chapter.

1.4 Interaction of wind with turbine dynamics

The interaction between wind fluctuations and the turbine is very complex and a thorough model of the turbine, generator and control system is needed for simulating the influence of wind turbulence in power output (Karaki et al., 2002; Vilar-Moreno, 2003). The control scheme and its optimized parameters are proprietary and difficult to obtain from manufacturers and complex to induce from measurements usually available.

The turbine and micro-meteorological dynamics transform the combination of periodic and random wind variations into stochastic fluctuations in the power. These variations can be divided into equivalent wind variations and almost periodic events such as vibration, blade positions, etc. Turbulence, turbine wakes, gusts... are highly random and do not show a definite frequency (Sørensen et al., 2002; Sørensen et al., 2008). Non-cyclic power variations are usually regarded as the outcome of the random component of the wind. They concern the control (short term prediction) and the forecast (long horizon prediction). Artificial Intelligence techniques and advanced filtering have been used for forecasting. Power fluctuations of frequency around 8 Hz can eventually produce flicker in very weak networks (Thiringer et al., 2004; Amaris & Usaola, 1997).

Both current and power can be measured directly, they can be statistically characterized and they are directly related to power quality. Current is transformed and its level depends on transformer ratio and actual network voltage. In contrast, power flows along transformers and networks without being altered except for some efficiency losses in the elements. That is why linearized power flows in the frequency domain are used in this chapter for characterizing experimentally the electrical behaviour of wind turbines.

1.5 Major difficulties in the fluctuation characterization

A priori estimation of power fluctuations requires thorough models of the wind turbines and turbulence. However, an empirical analysis is much simpler since distinct fluctuation sources usually present characteristic frequencies or some trend in the spectrum. In the following sections, a phenomenological and pragmatic approach will be applied to draw some conclusions and to extrapolate results from empirical studies to general cases.

The tower shadow, wind shear, rotor asymmetry and unbalance, blade misalignments produce a torque modulation dependent on turbine angle. This torque is filtered by turbine dynamics and the influence in output power can be complex. The signals cannot be considered truly periodic because neither the characteristic frequencies are constant (rotor speed is not constant and hence, the frequency of fluctuations induced by rotational effects) nor the frequencies are harmonically related. Some frequencies cannot be expressed as multiple of the others because the tower, blades and cinematic train present characteristic structural resonance frequencies different from the blade passing the tower frequency, f_{blade}. Moreover, turbine control, electric generator and power electronics introduce oscillations at other frequencies.

The turbulence adds a "coloured noise" overimposed to the former oscillatory modes, modulating cyclic vibrations and influencing rotor speed. The actual power is the outcome of many processes that interact and the analysis in the frequency domain is a simplifying approximation of a system driven by stochastic differential equations.

The first problem when analyzing power variations is that the contributions from rotor sampling, vibration modes and turbulence-driven variations are aggregated.

The second difficulty is the fact that frequencies of almost cyclic contributions are neither fixed nor are they multiple. Fourier coefficients are defined for periodic signals, but the sum of periodic components not harmonically related is no longer periodic.

The third difficulty is that frequencies of contributions are overlapped. Fortunately, characteristic frequencies (resonance and blade frequencies and its harmonics) have narrow margins for given operational conditions, producing peaks in the spectrum where one contribution usually predominates over the rest.

The forth difficulty is the turbulence, that introduces a non-periodic stochastic behaviour interacting with periodic signals. Different mathematical tools are customarily used for periodic and stochastic signals, increasing the difficulty of the analysis of these mixed-type signals.

The cyclic fluctuations of the turbine power can be considered in the *fraction-of-time* probability framework as the sum of sets of signals with different periods with additive stationary coloured noise and, hence, almost cyclostationary (Gardner et al, 2006). Since wind power is formed by the superposition of several almost cyclostationary signals whose periods are not harmonically related, wind power is *polycyclostationary*.

2. Mathematical framework and notation

2.1 Model assumptions

According to (Cidrás et al., 2002), voltage drops can only induce synchronized power fluctuations in a weak electrical network with a very steady and a very uniformly distributed wind. Most grid codes have been modified to minimize the simultaneous loss of generation during special events such as breaker tripping, grid transients, sudden voltages changes, etc. Except during the previous events, the synchronization of power fluctuations from a cluster of turbines is primarily due to wind variations that are slow enough to affect several turbines inside a wind farm.

Experimental measurements have corroborated that blade synchronisation is unusual. In addition, fluctuations due to turbine vibration, dynamics and control can be considered statistically independent between turbines, whereas turbulence and weather dynamics are partially correlated. Fortunately, slow fluctuations can be linked to equivalent wind fluctuations through a quasi-static approximation based on the power curve of the turbines.
As an outcome, the total fluctuation from an area is best characterized as a stochastic signal even though the fluctuations from single turbines have strong cyclic components. In other words, the transformation of cyclic components into stochastic components eases the treatment of wind farm power fluctuations.
For convenience, the signal duration will be considered short enough to be stationary (atmospheric dynamics will be supposed not to change considerably during the sample). Therefore, the average power (which corresponds to the zero frequency component of the sample) will be considered a known parameter.

a) Stochastic spectral phasor density of the active power

If $P(t)$ is the active power recorded in $0 \le t \le T$, its conventional Fourier transform, denoted by $\mathfrak{F}$, is scaled by a factor $1/\sqrt{T}$ to achieve an spectral measure whose main statistical properties do not depend on the sample duration T.

$$\vec{P}(f) \equiv P(f)\, e^{j\varphi(f)} \equiv \frac{1}{\sqrt{T}} \int_0^T P(t)\, e^{-j2\pi f\, t} dt = \frac{1}{\sqrt{T}} \mathfrak{F}\{P(t)\} \tag{1}$$

The factor $1/\sqrt{T}$ is between unity -used for pulses and signals of bounded energy- and $1/T$ -used in the Fourier coefficients of pure periodic signals-.
Fortunately, definition (1) has the advantage that the variance of $\vec{P}(f)$ is the two-sided auto spectral density, $\left\langle |\vec{P}(f)|^2 \right\rangle = PSD_P(f)$, which is independent of sample length T and it characterizes the process. $\vec{P}(f)$ will be referred as stochastic spectral phasor density of the active power or just the (stochastic) phasor for short.
Historically, the term "power spectral density" was coined when the signal analyzed $P(t)$ was the electric or magnetic field of a wave or the voltage output of an antenna connected to a resistor R. The power transferred to the load R at frequencies between $f\text{-}\Delta f/2$ and $f+\Delta f/2$ was $2\Delta f\cdot PSD_P(f)/R$ -that is proportional to $PSD_P(f)$ and the frequency interval. If $P(t)$ is the electric or magnetic field of a wave, then the power density at frequency f of that wave is also proportional to $\Delta f\cdot PSD_P(f)$.
In this chapter, $P(t)$ represents the power output of a turbine or a wind farm. The root mean square value (RMS for short) of power fluctuations at frequencies between $f\text{-}\Delta f/2$ and $f+\Delta f/2$ is $|\vec{P}(f)|\cdot\sqrt{2\cdot\Delta f}$. Power variance inside the previous frequency range is $PSD_P(f)\cdot\Delta f$. Hence, $PSD_P(f)$ in this chapter does not represent a power spectral density and this term can lead to misinterpretations. Therefore, $PSD_P(f)$ will be referred in this chapter as the *auto* spectral density although the acronym *PSD* (from Power Spectral Density) is maintained because it is widespread. Sometimes $PSD_P(f)$ will be replaced by $\sigma_P^2(f)$ to emphasize that it represents the *variance* spectral density of signal P at frequency f.
Fig. 3. shows the estimated *PSD* from 13 minute operation of a squirrel cage induction generator (SCIG) directly coupled to the grid (a portion of the original data is plotted in Fig. 1). The original auto spectrum is plotted in grey whereas the estimated *PSD* is in thin black

(linearly averaged periodogram in squared effective watts of real power per hertz). The trend is plotted in thick red, the accumulated variance is plotted in blue, and the tower shadow frequency is marked in yellow.

The instantaneous output of a wind farm or turbine can be expressed in frequency components using stochastic spectral phasor densities. As aforementioned, experimental measurements indicate that wind power nature is basically stochastic with noticeable fluctuating periodic components.

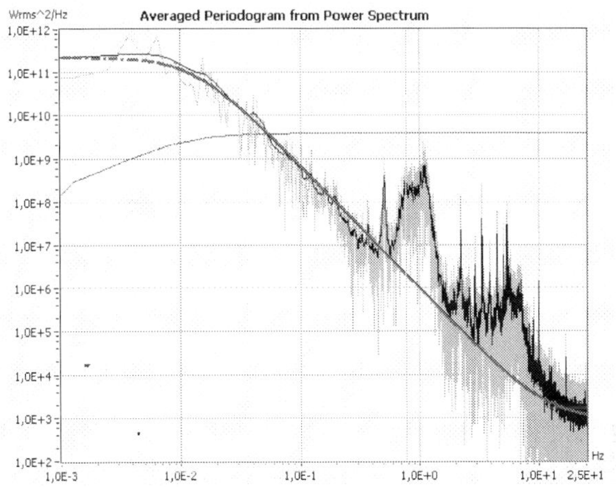

Fig. 3. $PSD_{P}(f)$ parameterization of active power of a 750 kW wind turbine for wind speeds around 6,7 m/s (average power 190 kW) computed from 13 minute data.

The signal in the time domain can be computed from the inverse Fourier transform:

$$P(t)=\sqrt{T}\int_{-\infty}^{\infty}\vec{P}(f)\,e^{j2\pi f t}df \underset{\vec{P}(f)=\vec{P}^{*}(-f)}{=} 2\sqrt{T}\int_{0}^{\infty}P(f)\cos\left[2\pi f\,t+\varphi(f)\right]df \qquad (2)$$

An analogue relation can be derived for reactive power and wind, both for continuous and discrete time. Standard FFT algorithms use two sided spectra, with negative frequencies in the last half of the output vector. Thus, calculus will be based on two-sided spectra unless otherwise stated, as in (2). In real signals, the negative frequency components are the complex conjugate of the positive one and a ½ scale factor may be applied to transform one to two-sided magnitudes.

b) Spectral power balance in a wind farm

Fluctuations at the point of common coupling (PCC) of the wind farm can be obtained from power balance equations for the average complex power of the wind farm.

Neglecting the increase in power losses in the grid due to fluctuating generation, the sum of oscillating power from the turbines equals the farm output undulation. Therefore, the complex sum of the frequency components of each turbine $\vec{P}_{turbine\ i}(f)$ totals the approximate farm output, $\vec{P}_{farm}(f)$.

$$\vec{P}_{farm}(f) \cong \sum_{i=1}^{N_{turbines}} \frac{\partial P_{farm}}{\partial P_{turbine\ i}} \vec{P}_{turbine\ i}(f) \approx \sum_{i=1}^{N_{turbines}} \eta_i \vec{P}_{turbine\ i}(f) = \sum_{i=1}^{N_{turbines}} \eta_i P_{turbine\ i}(f) e^{j\varphi_i(f)} \quad (3)$$

For usual wind farm configurations, total active losses at full power are less than 2% and reactive losses are less than 20%, showing a quadratic behaviour with generation level (Mur-Amada & Comech-Moreno, 2006). A small-signal model of power losses due to fluctuations inside the wind farm can be derived (Kundur et al. 1994), but since they are expected to be up to 2% of the fluctuation, the increase of power losses due to oscillations can be neglected in the first instance. A small signal model can be used to take into account network losses multiplying the turbine phasors in (3) by marginal efficiency factors $\eta_i = \partial P_{farm} / \partial P_{turbine\ i}$ estimated from power flows with small variations from the mean values using methodologies as the point-estimate method (Su, 2005; Stefopoulos et al., 2005). Typical values of η_i are about 98% for active power and about 85% for reactive power. In some expressions of this chapter, the efficiency has been set to 100% for clarity in the formulas.

In some applications, we encounter a random signal that is composed of the sum of several random sinusoidal signals, e.g., multipath fading in communication channels, clutter and target cross section in radars, interference in communication systems, wave propagation in random media and channels, laser speckle patterns and light scattering and summation of random current harmonics such as the ones produced by high frequency power converters of wind turbines (Baghzouz et al., 2002; Tentzerakis & Papathanassiou, 2007).

Any random sinusoidal signal can be considered as a random phasor, i.e., a vector with random length and angle. In this way, the sum of random sinusoidal signals is transformed into the sum of 2-D random vectors. So, irrespective of the type of application, we encounter the following general mathematical problem: there are vectors with lengths $P_i = |\vec{P}_i|$ and angles $\varphi_i = Arg(\vec{P}_i)$, in polar coordinates, where P_i and φ_i are random variables, as in (3) and Fig. 4. It is desired to obtain the probability density function (pdf) of the modulus and argument of the resulting vector. A comprehensive literature survey on the sum of random vectors can be obtained from (Abdi, 2000).

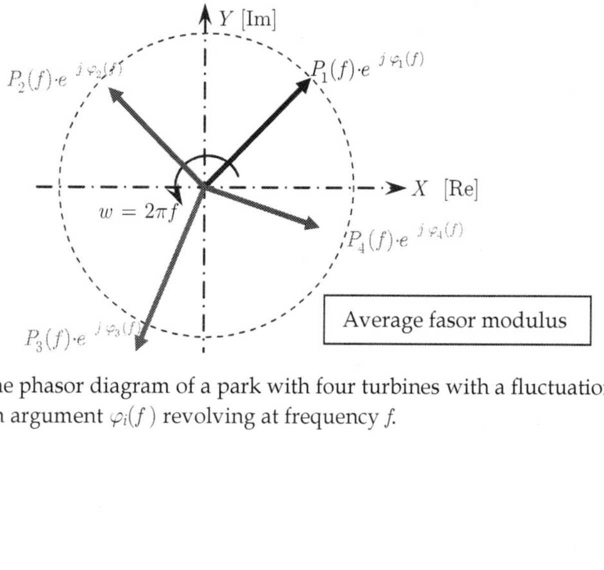

Fig. 4. Model of the phasor diagram of a park with four turbines with a fluctuation level $P_i(f)$ and random argument $\varphi_i(f)$ revolving at frequency f.

The vector sum of the four phasor in Fig. 4 is another random phasor corresponding to the farm phasor, provided the farm network losses are negligible. If some conditions are met, then the farm phasor can be modelled as a complex normal variable. In that case, the phasor amplitude has a Rayleigh distribution. The frequency $f = 0$ corresponds to the special case of the average signal value during the sample.

c) One and two sided spectra notation

One or two sided spectra are consistent -provided all values refer exclusively either to one or to two side spectra. Most differences do appear in integral or summation formulas - if two-sided spectra is used, a factor 2 may appear in some formulas and the integration limits may change from only positive frequencies to positive and negative frequencies.
One-sided quantities are noted in this chapter with a + in the superscript unless the differentiation between one and two sided spectra is not meaningful. For example, the one-sided stochastic spectral phasor density of the active power at frequency f is:

$$\left|\vec{P}^{+}(f)\right| = \left|\vec{P}(f)\right| + \left|\vec{P}(-f)\right| = 2\left|\vec{P}(f)\right| \tag{4}$$

In plain words, the one-sided density is twice the two-sided density. For convenience, most formulas in this chapter are referred to two-sided values.

d) Case study

Fig. 5 to Fig 8 show the power fluctuations of a wind farm composed by 27 wind turbines of 600 kW with variable resistance induction generator from VESTAS (Mur-Amada, 2009). The data-logger recorded signals either at a single turbine or at the substation. In either case, wind speed from the meteorological mast of the wind farm was also recorded.
The record analyzed in this subsection corresponds to date 26/2/1999 and time 13:52:53 to 14:07:30 (about 14:37 minutes). The average blade frequency in the turbines was $f_{blade} \approx 1{,}48 \pm 0{,}03$ Hz during the interval. The wind speed, measured in a meteorological mast at 40 m above the surface with a propeller anemometer, was $U_{wind} = 7{,}6$ m/s $\pm 2{,}0$ m/s (expanded uncertainty).
The oscillations due to rotor position in Fig. 5 are not evident since the total power is the sum of the power from 26 unsynchronized wind turbines minus losses in the farm network. Fig. 6 shows a rich dynamic behaviour of the active power output, where the modulation and high frequency oscillations are superimposed to the fundamental oscillation.

3. Asymptotic properties of the wind farm spectrum

The fluctuations of a group of turbines can be divided into the correlated and the uncorrelated components.
On the one hand, slow fluctuations ($f < 10^{-3}$ Hz) are mainly due to meteorological dynamics and they are widely correlated, both spatially and temporally. Slow fluctuations in power output of nearby farms are quite correlated and wind forecast models try to predict them to optimize power dispatch.
On the other hand, fast wind speed fluctuations are mainly due to turbulence and microsite dynamics (Kaimal, 1978). They are local in time and space and they can affect turbine control and cause flicker (Martins et al., 2006). Tower shadow is usually the most noticeable fluctuation of a turbine output power. It has a definite frequency and, if the blades of all turbines of an area became eventually synchronized, it could be a power quality issue.

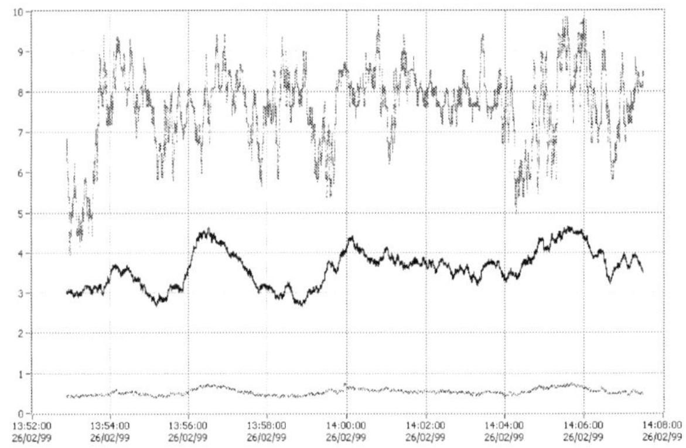

Fig. 5. Time series (from top to bottom) of the active power P [MW] (in black), wind speed U_{wind} [m/s] at 40 m in the met mast (in red) and reactive power Q [MVAr] (in dashed green).

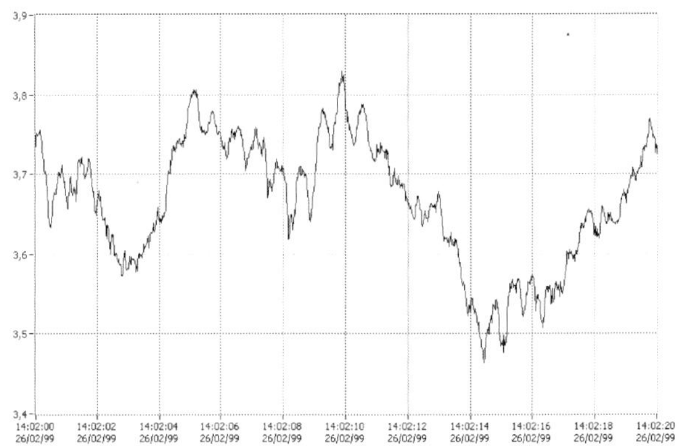

Fig. 6. Detail of the wind farm active power during 20 s at the wind farm.

The phase $\varphi_i(f)$ implies the use of a time reference. Since fluctuations are random events, there is not an unequivocal time reference to be used as angle reference. Since fluctuations can happen at any time with the same probability -there is no preferred angle $\varphi_i(f)$–, the phasor angles are random variables uniformly distributed in $[-\pi,+\pi]$ (i.e., the system exhibits circular symmetry and the stochastic process is cyclostationary). Therefore, the relevant information contained in $\varphi_i(f)$ is the relative angle difference among the turbines of the farm (Li et al., 2007) in the range $[-\pi,+\pi]$, which is linked to the time lag among fluctuations at the turbines.

The central limit for the sum of phasors is a fair approximation with 8 or more turbines and Gaussian process properties are applicable. Therefore, the wind farm spectrum converges asymptotically to a complex normal distribution, denoted by $\mathbb{C}N\left(0, \sigma_{Pfarm}(f)\right)$. In other words, $\mathrm{Re}[\vec{P}^{+}_{farm}(f)]$ and $\mathrm{Im}[\vec{P}^{+}_{farm}(f)]$ are independent random variables with normal distribution.

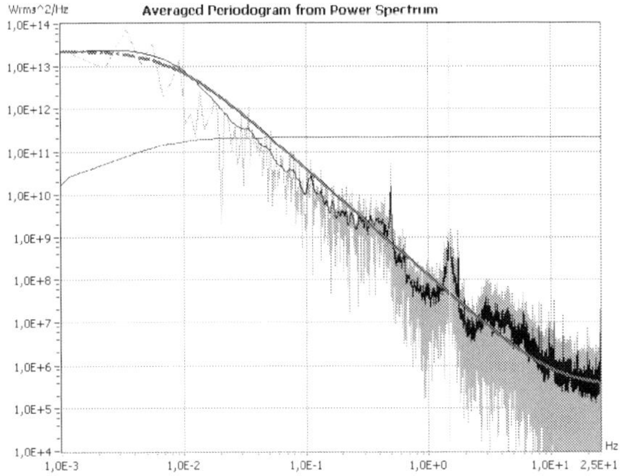

Fig. 7. $PSD_P+(f)$ parameterization of real power of a wind farm for wind speeds around 7,6 m/s (average power 3,6 MW) computed from data of Fig. 5.

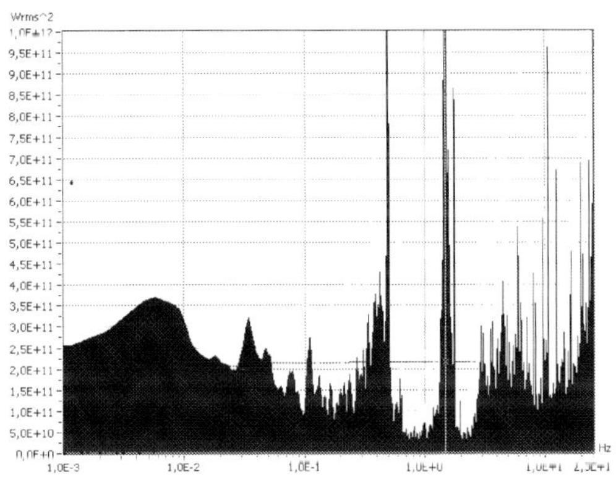

Fig. 8. Contribution of each frequency to the variance of power computed from Fig. 5 (the area bellow $f{\cdot}PSD_P+(f)$ in semi-logarithmic axis is the variance of power).

$$\vec{P}^{+}_{farm}(f) \sim \mathbb{C}N\left(0, \sigma_{farm}(f)\right) \tag{5}$$

Thus, the one-sided amplitude density of fluctuations at frequency f from N turbines, $\left|\vec{P}^{+}_{farm}(f)\right|$, is a Rayleigh distribution of scale parameter $\sigma_{Pfarm}(f) = \langle|\vec{P}^{+}_{farm}(f)|\rangle\sqrt{2/\pi}$, where angle brackets $\langle\bullet\rangle$ denotes averaging. In other words, the mean of $\left|\vec{P}^{+}_{farm}(f)\right|$ is $\langle|\vec{P}^{+}_{farm}(f)|\rangle = \sqrt{\pi/2}\,\sigma_{Pfarm}(f)$ where $\sigma_{Pfarm}(f)$ is the *RMS* value of the phasor projection. The *RMS* value of the phasor projection $\sigma_{Pfarm}(f)$ is also related to the one and two sided *PSD* of the active power:

$$\sigma_{Pfarm}(f) = \sqrt{2PSD_{Pfarm}(f)} = \sqrt{PSD^{+}_{Pfarm}(f)} \quad (6)$$

Put into words, the phasor density of the oscillation, $\left|\vec{P}^{+}_{Pfarm}(f)\right|$, has a Rayleigh distribution of scale parameter $\sigma_{Pfarm}(f)$ equal to the square root of the auto spectral density (the equivalent is also hold for two-sided values). The mean phasor density modulus is:

$$\langle|\vec{P}^{+}_{Pfarm}(f)|\rangle \underset{Rayleigh(\sigma_{Pfarm}(f))}{=} \sqrt{\frac{\pi}{2}}\sigma_{Pfarm}(f) \quad (7)$$

For convenience, effective values are usually used instead of amplitude. The effective value of a sinusoid (or its root mean square value, RMS for short) is the amplitude divided by √2. Thus, the average quadratic value of the fluctuation of a wind farm at frequency *f* is:

$$\left\langle\left|\vec{P}^{+}_{Pfarm}(f)/\sqrt{2}\right|^2\right\rangle = \left\langle\left|\vec{P}^{+}_{Pfarm}(f)\right|^2\right\rangle/2 \underset{Rayleigh[\sigma_N(f)]}{=} \sigma^2_{Pfarm}(f) = PSD^{+}_{Pfarm}(f) \quad (8)$$

If the active power of the turbine cluster is filtered with an ideal *narrowband filter* tuned at frequency f and bandwidth Δf, then *the average effective value* of the filtered signal is $\sigma_{Pfarm}(f)\sqrt{\Delta f}$ and the *average amplitude* of the oscillations is $\langle|\vec{P}^{+}_{farm}(f)|\rangle\cdot\sqrt{\Delta f} = \sigma_{Pfarm}(f)\sqrt{\Delta f\cdot\pi/2}$. The instantaneous value of the filtered signal $P_{Pfarm,f,\Delta f}(t)$ is the projection of the phasor $\vec{P}^{+}_{farm}(f)\cdot e^{j2\pi f\,t}\sqrt{\Delta f}$ in the real axis. The instantaneous value of the square of the filtered signal, $P^2_{farm,f,\Delta f}(t)$, is an exponential random variable of parameter $\lambda=[\sigma^2_{farm}(f)\Delta f]^{-1}$ and its mean value is:

$$\left\langle P^2_{farm,f,\Delta f}(t)\right\rangle \underset{Exp\ distribution}{=} \lambda = \sigma^2_{Pfarm}(f)\Delta f \quad (9)$$

For a continuous *PSD*, the expected variance of the instantaneous power output during a time interval T is the integral of $\sigma_{Pfarm}(f)$ between $\Delta f = 1/T$ and the grid frequency, according to Parseval's theorem (notice that the factor 1/2 must be changed into 2 if two-sided phasors densities are used):

$$\left\langle P^2_{farm}(t)\right\rangle = \frac{1}{2}\left\langle \int_{1/T}^{f_{grid}} |\vec{P}^{+2}_{farm}(f)|\, df\right\rangle = \frac{1}{2}\int_{1/T}^{f_{grid}} \langle|\vec{P}^{+2}_{farm}(f)|\rangle\, df = \int_{1/T}^{f_{grid}} \sigma^2_{farm}(f)\, df \quad (10)$$

In fact, data is sampled and the expected variance of the wind farm power of duration T can be computed through the discrete version of (10), where the frequency step is $\Delta f = 1/T$ and the time step is $\Delta t = T/m$:

$$\left\langle P^2_{farm}(t)\right\rangle = \frac{1}{2}\left\langle \sum_{k=1}^{m-1} |\vec{P}^{+2}_{farm}(k\Delta f)|\Delta f\right\rangle = \frac{1}{2}\sum_{k=1}^{m-1} \langle|\vec{P}^{+2}_{farm}(k\Delta f)|\rangle\Delta f = \sum_{k=1}^{m-1}\sigma^2_{Pfarm}(k\Delta f)\Delta f \quad (11)$$

If a fast Fourier transform is used as a narrowband filter, an estimate of $\sigma^2_{Pfarm}(f)$ for $f = k\,\Delta f$ is $2\Delta f \cdot \langle |FFT_k\{P_{farm}(i\Delta t)\}|^2\rangle$. In fact, the factor $2\Delta f$ may vary according to the normalisation factor included in the *FFT*, which depends on the software used. Usually, some type of smoothing or averaging is applied to obtain a consistent estimate, as in Bartlett or Welch methods (Press et al., 2007).

The distribution of $\left\langle P^2_{farm}(t)\right\rangle$ can be derived in the time or in the frequency domain. If the process is normal, then the modulus and phase of $\vec{P}^+_{farm}(f_k)$ are not linearly correlated at different frequencies f_k. Then $\left\langle P^2_{farm}(t)\right\rangle$ is the sum in (11) or the integration in (10) of independent Exponential random variables that converges to a normal distribution with mean $\left\langle P^2_{farm}(t)\right\rangle$ and standard deviation $\sqrt{2}\left\langle P^2_{farm}(t)\right\rangle$.

In farms with a few turbines, the signal can show a noticeable periodic fluctuation shape and the auto spectral density $\sigma^2_{Pfarm}(f)$ can be correlated at some frequencies. These features can be discovered through the bispectrum analysis. In such cases, $\left\langle P^2_{farm}(t)\right\rangle$ can be computed with the algorithm proposed in (Alouini et al., 2001).

4. Sum of partially correlated phasor densities of power from several turbines

4.1 Sum of fully correlated and fully uncorrelated spectral components

If turbine fluctuations at frequency f of a wind farm with N turbines are completely synchronized, all the phases have the same value $\varphi(f)$ and the modulus of fully correlated fluctuations $|\vec{P}^+_{i,\,corr}(f)|$ sum arithmetically:

$$|\vec{P}^+_{farm,\,corr}(f)| = \left|\sum_{i=1}^{N}\eta_i\vec{P}^+_{i,\,corr}(f)\right| = \sum_{i=1}^{N}|\eta_i\vec{P}^+_{i,\,corr}(f)| \tag{12}$$

If there is no synchronization at all, the fluctuation angles $\varphi_i(f)$ at the turbines are stochastically independent. Since $\vec{P}_{i,\,uncorr}(f)$ has a random argument, its sum across the wind farm will partially cancel and inequality (13) holds true.

$$|\vec{P}^+_{farm,\,uncorr}(f)| = \left|\sum_{i=1}^{N}\eta_i\vec{P}^+_{i,\,uncorr}(f)\right| < \sum_{i=1}^{N}\eta_i|\vec{P}^+_{i,\,uncorr}(f)| \tag{13}$$

This approach remarks that correlated fluctuations adds arithmetically and they can be an issue for the network operation whereas uncorrelated fluctuations diminish in relative terms when considering many turbines (even if they are very noticeable at turbine terminals).

A) Sum of uncorrelated fluctuations

The fluctuation of power output of the farm is the sum of contributions from many turbines (3), which are mainly uncorrelated at frequencies higher than a tenth of Hertz.

The sum of N independent phasors of random angle of N equal turbines in the farm converges asymptotically to a complex Gaussian distribution, $\vec{P}_{farm}(f) \sim \mathcal{CN}[0, \sigma_{Pfarm}(f)]$, of null mean and standard deviation $\sigma_{farm}(f) = \sqrt{\eta N}\sigma_1(f)$, where $\sigma_1(f)$ is the mean *RMS* fluctuation at a single turbine at frequency f and η is the average efficiency of the farm network. To be precise, the variance $\sigma_1{}^2(f)$ is half the mean squared fluctuation amplitude at frequency f, $\sigma_1^2(f) = \frac{1}{2}\left\langle \left|\vec{P}_{turbine\ i}(f)\right|^2 \right\rangle = \left\langle \mathrm{Re}^2\left[\vec{P}_{turbine\ i}(f)\right]\right\rangle = \left\langle \mathrm{Im}^2\left[\vec{P}_{turbine\ i}(f)\right]\right\rangle$. Therefore, the real and imaginary phasor components $\mathrm{Re}[\vec{P}_{farm}(f)]$ and $\mathrm{Im}[\vec{P}_{farm}(f)]$ are independent real Gaussian random variables of standard deviation $\sigma_{Pfarm}(f)$ and null mean since phasor argument is uniformly distributed in $[-\pi,+\pi]$. Moreover, the phasor modulus $\left|\vec{P}_{farm}(f)\right|$ has $Rayleigh[\sigma_{Pfarm}(f)]$ distribution. The double-sided power spectrum $\left|\vec{P}_{farm}(f)\right|^2$ is an $Exponential\left[\lambda = \frac{1}{2}\sigma_{Pfarm}^{-2}(f)\right]$ random vector of mean $\left\langle \left|\vec{P}_{farm}(f)\right|^2 \right\rangle = 2\sigma_{Pfarm}^2(f) = \frac{1}{2} PSD_{Pfarm}(f)$ (Cavers, 2003).

The estimate from the periodogram is the moving average of $N_{aver.}$ exponential random variables corresponding to adjacent frequencies in the power spectrum vector. The estimate is a Gamma random variable. If the *PSD* is sensibly constant on $N_{aver}\Delta f$ bandwidth, then the *PSD* estimate has the same mean as the original *PSD* and the standard deviation is $\sqrt{N_{aver.}}$ times smaller (i.e., the estimate has lower uncertainty at the cost of lower frequency resolution).

4.2 Sum of partially linearly correlated spectral components

Inside a farm, the turbines usually exhibit a similar behaviour for a given frequency f and the *PSD* of each turbine is expected to be fairly similar. However, the phase differences among turbines do vary with frequency. Slow meteorological variations affect all the turbines with negligible time lag, compared to characteristic time frame of weather systems (i.e., the phasors $\vec{P}_{turbine}(f)$ have the same phase). Turbulences with scales significantly smaller than the turbine distances have uncorrelated phases. Fluctuations due to rotor positions also show uncorrelated phases provided turbines are not synchronized.

$$\left\langle P_{turbine}^{+}(f)\right\rangle^2 = \left\langle P_{turb,corr}^{+}(f)\right\rangle^2 + \left\langle P_{turb,uncorr}^{+}(f)\right\rangle^2 \tag{14}$$

If the number of turbines N >4 and the correlation among turbines are linear, the central limit is a good approximation. The correlated and uncorrelated components sum quadratically and the following relation is applicable:

$$\left\langle \left|\vec{P}_{farm}^{+}(f)\right|^2 \right\rangle \approx (\eta N)^2 \left\langle \left|\vec{P}_{turb,\ corr}^{+}(f)\right|^2 \right\rangle + \eta N \left\langle \left|\vec{P}_{turb,\ uncorr}^{+}(f)\right|^2 \right\rangle \tag{15}$$

where N is the number of turbines in the farm (or in a group of close farms) and η is the average efficiency of the farm network (typical values are about 98% for active power and about 85% for reactive power). Since phasor densities sum quadratically, (14) and (15) are concisely expressed in terms of the *PSD* of correlated and uncorrelated components of phasor density:

$$PSD_{farm}(f) \approx (\eta N)^2\ PSD_{turb,\ corr}(f) + \eta N \cdot PSD_{turb,\ uncorr}(f) \tag{16}$$

$$PSD_{turb}(f) = PSD_{turb,\ corr}(f) + PSD_{turb,\ uncorr}(f) \tag{17}$$

The correlated components of the fluctuations are the main source of fluctuation in large clusters of turbines. The farm admittance $J(f)$ is the ratio of the mean fluctuation density of the farm, $\left\langle \left| \vec{P}_{farm}(f) \right| \right\rangle$, to the mean turbine fluctuation density, $\left\langle \left| P^+_{turbine}(f) \right| \right\rangle$.

$$J(f) = \frac{\left\langle \left| P^+_{farm}(f) \right| \right\rangle}{\left\langle \left| P^+_{turbine}(f) \right| \right\rangle} \approx \sqrt{\frac{PSD_{Pfarm}(f)}{PSD_{Pturbine}(f)}} \tag{18}$$

Note that the phase of the admittance $J(f)$ has been omitted since the phase lag between the oscillations at the cluster and at a turbine depend on its position inside the cluster. The admittance is analogous to the expected gain of the wind farm fluctuation respect the turbine expected fluctuation at frequency f (the ratio is referred to the mean values because both signals are stochastic processes).

Since turbine clusters are not negatively correlated, the following inequality is valid:

$$\sqrt{\eta N} \lesssim J(f) \lesssim \eta N \tag{19}$$

The squared modulus of the admittance $J(f)$ is conveniently estimated from the *PSD* of the turbine cluster and a representative turbine using the cross-correlation method and discarding phase information (Schwab et al., 2006):

$$J^2(f) = \frac{PSD_{Pfarm}(f)}{PSD_{Pturb}(f)} = (\eta N)^2 \frac{PSD_{turb,corr}(f)}{PSD_{turb}(f)} + \eta N \frac{PSD_{turb,uncorr}(f)}{PSD_{turb}(f)} \tag{20}$$

If the PSD of a representative turbine, $PSD_{Pturb}(f)$, and the PSD of the farm $PSD_{Pfarm}(f)$ are available, the components $PSD_{turb,corr}(f)$ and $PSD_{turb,uncorr}(f)$ can be estimated from (16) and (17) provided the behaviour of the turbines is similar.

At $f \ll 0{,}01$ Hz, fluctuations are mainly correlated due to slow weather dynamics, $PSD_{turb,uncorr}(f) \ll PSD_{turb,corr}(f)$, and the slow fluctuations scale proportionally $PSD_{Pfarm}(f) \approx (\eta N)^2 PSD_{turb,corr}(f)$. At $f > 0{,}01$ Hz, individual fluctuations are statistically independent, $PSD_{turb,uncorr}(f) \gg PSD_{turb,corr}(f)$, and fast fluctuations are partially attenuated, $PSD_{Pfarm}(f) \approx \eta N \cdot PSD_{turb,uncorr}(f)$.

An analogous procedure can be replicated to sum fluctuations of wind farms of a geographical area, obtaining the correlated $PSD_{farm,corr}(f)$ and uncorrelated $PSD_{farm,uncorr}(f)$ components. The main difference in the regional model -apart from the scattered spatial region and the different turbine models- is that wind farms must be normalized and an average farm model must be estimated for reference. Therefore, the average farm behaviour is a weighted average of individual farms with lower characteristic frequencies (Norgaard & Holttinen, 2004). Recall that if hourly or even slower fluctuations are studied, meteorological dynamics are dominant and other approaches are more suitable.

4.3 Estimation of wind farm power admittance from turbine coherence

The admittance can be deducted from the farm power balance (3) if the coherence among the turbine outputs is known. The system can be approximated by its second-order statistics

as a multivariate Gaussian process with spectral covariance matrix $\Xi_P(f)$. The elements of $\Xi_P(f)$ are the complex squared coherence at frequency f and at turbines i and j, noted as $\vec{\gamma}_{ij}(f)$. The efficiency of the power flow from the turbine i to the farm output can be expressed with the column vector $\eta_P = [\eta_1, \eta_2, ..., \eta_N]^T$, where ^{T} denotes transpose.
Therefore, the wind farm power admittance $J(f)$ is the sum of all the coherences, multiplied by the efficiency of the power flow:

$$J^2(f) \approx \left| \sum_{i=1}^{N} \sum_{j=1}^{N} \eta_i \eta_j \vec{\gamma}'_{ij}(f) \right| = \left| \eta_P^T \; \Xi_P(f) \; \eta_P \right| \tag{21}$$

The squared admittance for a wind farm with a grid layout of n_{long} columns separated d_{long} distance in the wind direction and n_{lat} rows separated d_{lat} distance perpendicular to the wind U_{wind} is:

$$J^2(f) \approx \eta^2 \sum_{i_1=1}^{n_{lat}} \sum_{i_2=1}^{n_{lat}} \sum_{j_1=1}^{n_{long}} \sum_{j_2=1}^{n_{long}} Cos\left[\frac{2\pi(j_2 - j_1)d_{long} f}{U_{wind}}\right] Exp\left[\frac{-f\sqrt{A_{lat}^2 (i_2 - i_1)^2 d_{lat}^2 + A_{long}^2 (j_2 - j_1)^2 d_{long}^2}}{U_{wind}}\right] \tag{22}$$

The admittance computed for Horns Rev offshore wind farm (with a layout similar to Fig. 10) is plotted in Fig. 9. According to (Sørensen et al., 2008), it has 80 wind turbines disposed in a grid of n_{lat} = 8 rows and n_{long} = 10 columns separated by seven diameters in each direction (d_{lat} = d_{long} = 560 m), high efficiency ($\eta \approx$ 100%), lateral coherence decay factor $A_{lat} \approx U_{wind}/$(2 m/s), longitudinal coherence decay factor $A_{long} \approx$ 4, wind direction aligned with the rows and $U_{wind} \approx$ 10 m/s wind speed.

4.4 Estimation of wind farm power admittance from the wind coherence

The wind farm admittance $J(f)$ can be approximated from the equivalent farm wind because the coherence of power and wind are similar (the transition frequency between correlated and uncorrelated behaviour is about 10^{-2} Hz for small wind farms). According to (Mur-Amada, 2009), the equivalent wind can be roughly approximated by a multivariate

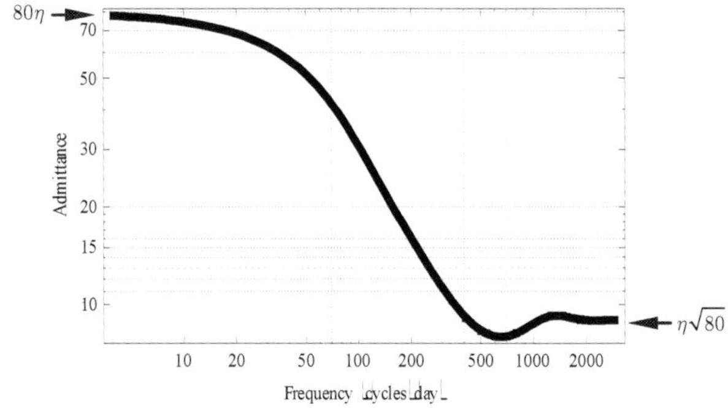

Fig. 9. Admittance for Horns Rev offshore wind farm for 10 m/s and wind direction aligned with the turbine rows.

Gaussian process with spectral covariance matrix $\Xi_{Ueq}(f)$. Its elements are the complex coherence of effective turbulence at frequency f and at turbines i and j, denoted by $\vec{\gamma}'_{ij}(f)$. In this case, the column vector $\eta_{Ueq} = [\eta'_1, \eta'_2, ..., \eta'_N]^T$ should be interpreted as the relative sensitivity of the farm power respect the equivalent wind in each turbine. Therefore, the wind farm power admittance $J(f)$ is the sum of the complex coherence of effective quadratic turbulence among turbines:

$$J^2(f) \approx \left| \sum_{i=1}^{N} \sum_{j=1}^{N} \eta'_i \eta'_j \vec{\gamma}'_{ij}(f) \right| = \left| \eta^T_{Ueq} \; \Xi_{Ueq}(f) \; \eta_{Ueq} \right| \tag{23}$$

For the rectangular region shown in Fig. 10, the admittance is:

$$J^2(f) \approx \eta N \left\{ 1 + (\eta N - 1) \, H^2(f) \right\} \tag{24}$$

where

$$H^2(f) = \frac{PSD_{Ueq,area}(f)}{PSD_{Ueq,turbine}(f)} = g\left(\frac{A_{lat} b \; f}{\langle U_{wind} \rangle} \right) \mathrm{Re}\left[g\left(\frac{(A_{long} + j2\pi) \; a \; f}{\langle U_{wind} \rangle} \right) \right] \tag{25}$$

$$g(x) = 2\left(-1 + e^{-x} + x \right) / \; x^2 \tag{26}$$

$\langle U_{wind} \rangle$ is the mean wind during the sample, η is the average sensitivity of the power respect the wind and a and b are the dimensions of the wind farm according to Fig. 10. The decay constants for lateral and longitudinal directions are, A_{long} and A_{lat}, respectively. For the Rutherford Appleton Laboratory, (Schlez & Infield, 1998) recommended $A_{long} \approx (15\pm5)$ $\sigma_{Uwind} / \langle U_{wind} \rangle$ and $A_{lat} \approx (17{,}5\pm5)$ $(m/s)^{-1}\sigma_{Uwind}$, where σ_{Uwind} is the standard deviation of the wind speed in m/s. IEC 61400-1 recommends $A \approx 12$; Frandsen (Frandsen et al., 2007) recommends $A \approx 5$ and Saranyasoontorn (Saranyasoontorn et al., 2004) recommends $A \approx 9{,}7$.

$H^2(f)$ is the quadratic coherence between the equivalent wind of the farm, relative to the turbine. $H(f)$ measures the correlation of the phase difference between the equivalent wind of the farm relative to the turbine at frequency f. If $H(f)$ is unity, the turbine phasors have

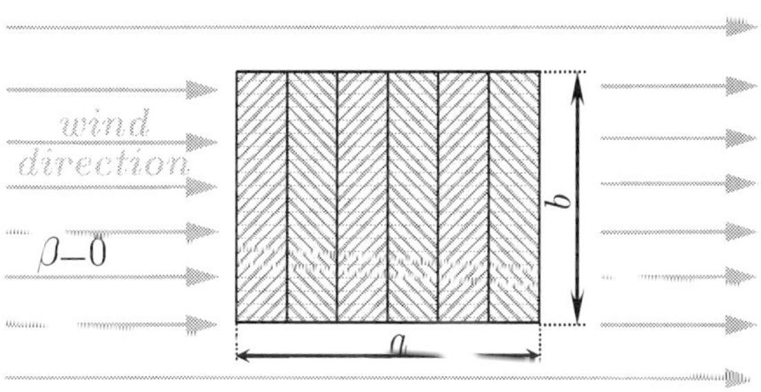

Fig. 10. Wind farm dimensions for the case of frontal wind direction.

the same angle and the turbine fluctuations are synchronized at that frequency. If $H(f)$ is zero, the phasors have uncorrelated arguments and hence, the turbine fluctuations are stochastically uncorrelated at that frequency. Hence, $H(f)$ is the correlation level at frequency f of the fluctuations among the turbines, measured from 0 to 1.
The transition frequency from correlated to uncorrelated fluctuations is obtained solving $H^2(f)$ =1/4. Thus, the cut-off frequency of narrow wind farms with $a \ll b$ is:

$$f_{cut,lat} = 6.83 \frac{\langle U_{wind} \rangle}{b A_{lat}} \tag{27}$$

In the Rutherford Appleton Laboratory (RAL), $A_{lat} \approx (17{,}5 \pm 5)(\mathrm{m/s})^{-1} \sigma_{Uwind}$ and hence $f_{cut,lat} \approx (0{,}42 \pm 0{,}12) \langle U_{wind} \rangle / (\sigma_{Uwind}\, b)$. A typical value of the turbulence intensity $\sigma_{Uwind} / \langle U_{wind} \rangle$ is around 0,12 and for such value $f_{cut,lat} \sim (3.5 \pm 1)/b$, where b is the lateral dimension of the area in meters. For a narrow farm of b = 3 km, the cut-off frequency is in the order of 1,16 mHz.
In Horns Rev wind farm, $A_{lat} = \langle U_{wind} \rangle / (2\ \mathrm{m/s})$ and hence $f_{cut,lat} \approx 13{,}66/b$, where b is a constant expressed in meters. For a wind farm of b = 3 km, the cut-off frequency is in the order of 4,5 mHz (about four times the estimation from RAL).
In RAL, $A_{long} \approx (15 \pm 5)\, \sigma_{Uwind} / \langle U_{wind} \rangle$. A typical value of the turbulence intensity $\sigma_{Uwind} / \langle U_{wind} \rangle$ is around 0,12 and for such value $A_{long} \approx (1{,}8 \pm 0{,}6)$.

$$f_{cut,long} \underset{A_{long} \sim 1{,}8}{=} 1{,}1839 \frac{\langle U_{wind} \rangle}{a\, A_{long}} \underset{A_{long} = 1.8}{=} 0.6577 \frac{\langle U_{wind} \rangle}{a} \tag{28}$$

For a significative wind speed of $\langle U_{wind} \rangle \sim 10$ m/s and a wind farm of a = 3 km longitudinal dimension, the cut-off frequency is in the order of 2,19 mHz.
In the Høvsøre wind farm, A_{long} = 4 (about twice the value from RAL). The cut-off frequency of a longitudinal area with A_{long} around 4 (dashed gray line in Fig. 11) is:

$$f_{cut,long} \underset{A_{long} \sim 4}{=} 2.7217 \frac{\langle U_{wind} \rangle}{a\, A_{long}} \underset{A_{long} = 4}{=} 0.6804 \frac{\langle U_{wind} \rangle}{a} \tag{29}$$

For a significative wind speed of $\langle U_{wind} \rangle \sim$ 10 m/s and a wind farm of a = 3 km longitudinal dimension, the cut-off frequency is in the order of 2,26 mHz.
In accordance with experimental measurements, turbulence fluctuations quicker than a few minutes are notably smoothed in the wind farm output. This relation is proportional to the dimensions of the area where the wind turbines are sited. That is, if the dimensions of the zone are doubled, the area is four times the original region and the cut-off frequencies are halved. In other words, *the smoothing of the aggregated wind is proportional to the longitudinal and lateral lengths* (and thus, related to the square root of the area if zone shape is maintained).
In sum, the lateral cut-off frequency is inversely proportional to the site parameters A_{lat} and the longitudinal cut-off frequency is only slightly dependent on A_{long}. Note that the longitudinal cut-off frequency show closer agreement for Høvsøre and RAL since it is dominated by frozen turbulence hypothesis.

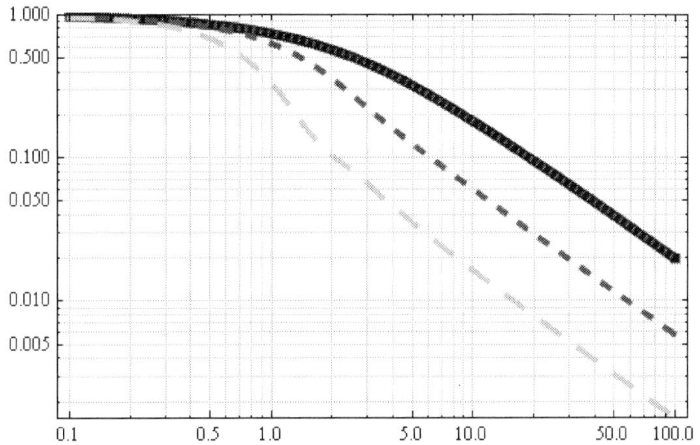

Fig. 11. Normalized ratio $H^2(f)$ for transversal $a \ll b$ (solid thick black line) and longitudinal $a \gg b$ areas (dashed dark gray line for $A_{long} = 4$, long dashed light gray line for $A_{long} = 1{,}8$). Horizontal axis is expressed in either longitudinal or lateral adimensional frequency $a\ A_{long}\ f / \langle U_{wind} \rangle$ or $b\ A_{lat}\ f / \langle U_{wind} \rangle$.

However, if transversal or longitudinal smoothing dominates, then the cut-off frequency is approximately the minimum of $f_{cut,lat}$ and $f_{cut,long}$. The system behaves as a first order system at frequencies above both cut-off frequencies, and similar to a ½ order system between $f_{cut,lat}$ and $f_{cut,long}$.

5. Case study: comparison of PSD of a wind farm with respect to one of its turbines during 12 minutes

A literature review on experimental data of power output PSD from wind turbines or wind farms can be found in (Mur-Amada & Bayod-Rujula, 2007), with a parameterization and analysis of the data from very different locations. (Apt, 2007) shows an interesting comparison of the spectrum of the wind power from a wide area.

In this sub-section, the analysis of a case based on (Mur-Amada, 2009) is presented. The similarity of the *PSD* at one turbine and at the overall output of a wind farm of 18 turbines is shown. If the fluctuations at every turbine are independent (i.e. the turbines behaves independently from each other), then the *PSD* of the wind farm is approximately the *PSD* of each turbine multiplied by the number of turbines and by the power flow efficiency.

Each turbine experiments different turbulence levels and wind averages, so a representative turbine should be selected. The time lag between the variations measured in the farm and in the turbine depends on the farm layout. The phase information has been discarded because the phase of ergodic stochastic processes do not contain statistical information.

Fig. 12 shows the power output of the wind farm and the scalled output of one turbine. Since the measured turbine is more exposed to the wind than others turbines, the ratio of the average power of the turbine to the farm is 14 (less than 18, the number of turbines in the farm). There is a clear reduction of the relative variability in the farm output and some slow oscillations between the turbine and the farm seem to be delayed. In fact, this section will show that the ratio of the fluctuations is about $\sqrt{18}$ because the measured fluctuations are mainly uncorrelated, the duration of the sample is relatively short (less than 12 minutes) and the wind does not show a noticeable trend during the sample.

If the turbines behave independently from each other and they are similar, then the *PSD* of the wind farm is the *PSD* of one turbine times the number of turbines in the farm and times a power efficiency factor. To test this hypothesis, the farm *PSD* is shown in solid black and the turbine *PSD* times 18 is in dashed green in Fig. 13, with good agreement.

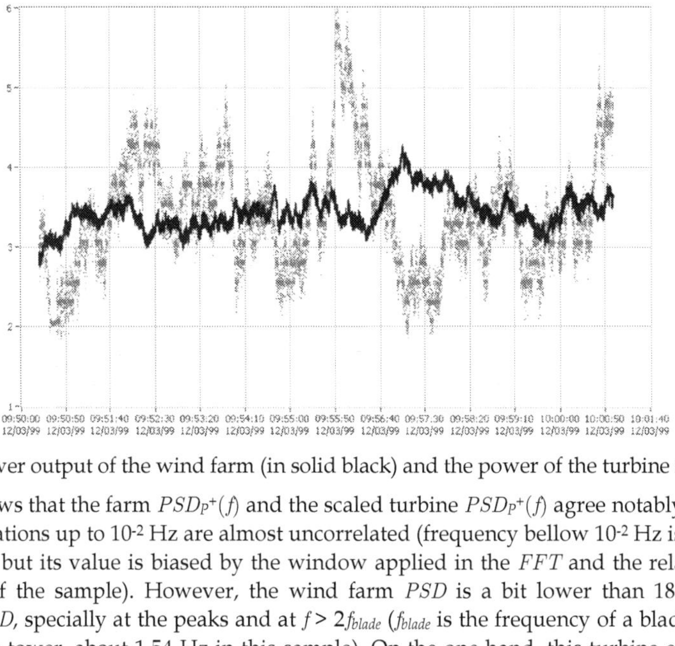

Fig. 12. Power output of the wind farm (in solid black) and the power of the turbine times 14.

Fig. 13 shows that the farm $PSD_{P^+}(f)$ and the scaled turbine $PSD_{P^+}(f)$ agree notably, showing that fluctuations up to 10^{-2} Hz are almost uncorrelated (frequency bellow 10^{-2} Hz is shown in the figure, but its value is biased by the window applied in the *FFT* and the relative short duration of the sample). However, the wind farm *PSD* is a bit lower than 18 times the turbine *PSD*, specially at the peaks and at $f > 2f_{blade}$ (f_{blade} is the frequency of a blade crossing the turbine tower, about 1,54 Hz in this sample). On the one hand, this turbine experiences more cyclic oscillations, partly due to a misalignment of the rotor bigger than the farm average. On the other hand, this turbine produced an average of $1/14^{th}$ of the wind farm power on the series #1 (see Fig. 12). This explain that *PSD* at $f > 2f_{blade}$ is primarily proportional to power output ratio (the farm *PSD* is 14 times the turbine *PSD*).

The real power admittance is shown in Fig. 14. The admittance is the ratio of the farm spectrum to the turbine spectrum of real power and it can be estimated as the square root of the *PSD* ratios. The level $\sqrt{18}$ has been added in dash-dotted red line to compare with the theoretical value of uncorrelated fluctuations.

In general terms, the assumption of uncorrelated fluctuations at frequencies higher than 10^{-2} Hz is valid: the admittance is approximately $\sqrt{18}$, the square root of the number of turbines in the farm. At $f > 2f_{blade}$, the admittance is more similar to $\sqrt{14}$ (the square root of the farm power divided by the turbine power). At $f < 0{,}02$ Hz, the admittance starts drifting from $\sqrt{18}$, indicating that oscillations at very low frequency are somewhat correlated.

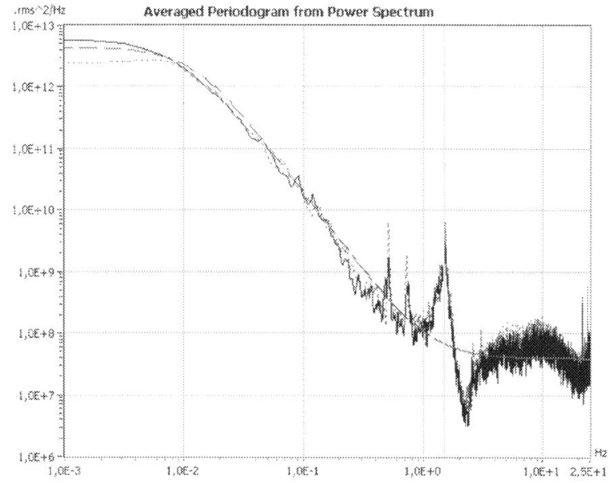

Fig. 13. $PSD_{Pfarm}^{+}(f)$ of a wind farm (in solid black) and $PSD_{Pturbine}^{+}(f)$ of one of its 648 kW turbines times 18 (in dashed green), for time series #1.

There is a peak in Fig. 14 at 2 Hz < f < 2,5 Hz. The analyzed turbine may have comparative less fluctuations in such range than the other turbines in the farm (the measured turbine may have better adjusted rotor and blades, while others turbines may suffer from more vibration effects). But other feasible reason is a higher correlation degree between the turbines at such frequency band, probably induced by turbine control or voltage variations.

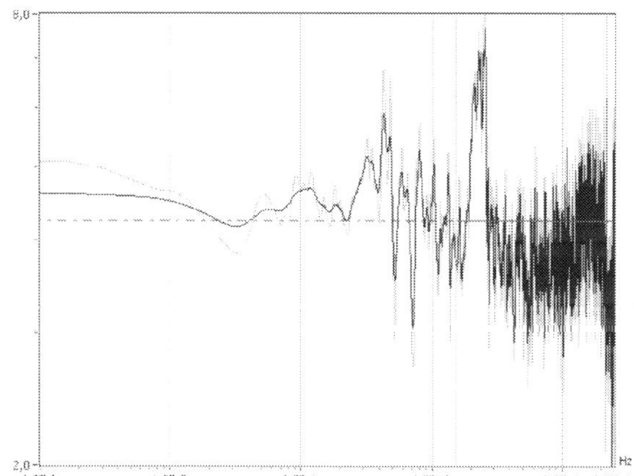

Fig. 14. Admittance of the active power (ratio of the farm *PSD* to the turbine *PSD*).

In short, real power oscillations quicker than one minute can be considered independent among turbines of a wind farm because the *PSD* due to fast turbulence and rotational effects scales proportionally to the number of turbines.

The former section has analyzed values logged with high time resolution (each grid cycle, 20 ms) but the duration was relatively short (a bit more than 10 minutes) due to storage limitations in the recording system. Ten-minute records with 20 ms time resolution allow studying fluctuations with durations between some tenths of second up to one minute However, this duration is insufficient for analyzing wind farm dynamics slower than 0.016 Hz with acceptable uncertainty.

6. Case study: comparison of PSD of a wind farm with respect to one of its turbines during a day

In order to study the behaviour of fluctuations slower than one minute, the next section will analyze the mean power of each second during a day. Daily records with one second time resolution allow to study the fluctuations with durations from a few seconds up to an hour.

Overall, the transition frequency from uncorrelated to correlated fluctuations is mild and, in fact, the ratio $PSD_{farm}(f)/PSD_{turbine}(f)$ depends noticeably on atmospheric conditions and it varies from one wind farm to another. This is one of the reasons why the values of the coherence decay factors A_{long} and A_{lat} may vary twofold among different sources.

At higher frequencies, the control and generator technology influences greatly the smoothness of the power delivery. At low frequencies and under rated power, the variability is mainly due to the wind because any turbine tries to extract the maximum amount of power from the wind, regardless of their technology. During full power generation, the fluctuations have smaller amplitude and higher frequency.

The case presented in this section corresponds to low/mid wind speed, since this range presents bigger fluctuations. The wind direction does not present big deviations during the day and the atmospheric conditions can be considered similar during all the day.

For clarity, the turbine and the farm is generating bellow rated power during all the day presented in this sections, without null, maximum power or unavailability periods. These operating conditions present quite different features, and each functioning mode should be treated differently. Moreover, some intermittent power delivery may occur during the transition from one operation condition to another, and this event should be treated as a transient. In fact, this chapter is limited to the analysis of continuous operation, without considering transitory events (such features can be better studied with other tools).

6.1 Daily spectrograms

The PSD in the *fraction-of-time* probability framework is the long term average of auto spectrum density and it characterizes the behaviour of stochastically stationary systems. The spectrogram shows the spectrum evolution and the stationarity of signals can be tested with it. Every spectrogram column can be thought as the power spectrum of a small signal sample. Therefore, the PSD in the classical stochastic framework is the ensemble average of the power spectrums. For stationary systems, the classical and the *fraction-of-time* approaches are equivalent.

The analysis has been performed using the spectrogram of the active power. The frequency band is between 0,5 Hz (fluctuations of 2 second of duration, corresponding to $8{,}4 \cdot 10^5$ cycles/day) and 6 cycles/day (fluctuations of 4 hours of duration).

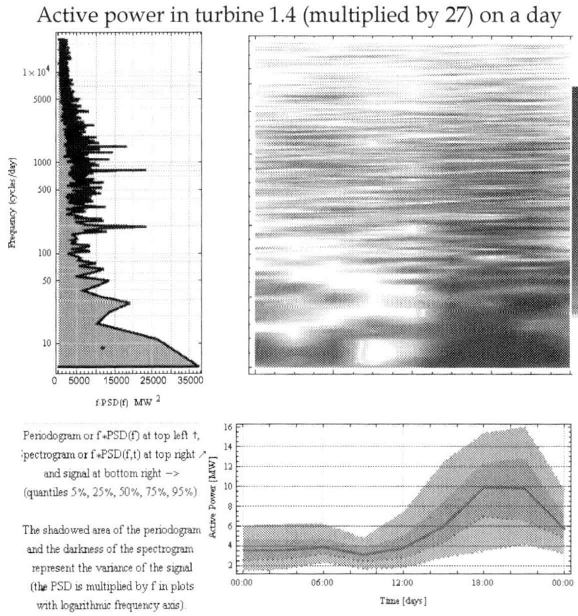

Fig. 15. Spectrogram of the real power [MW] at a turbine (times the turbines in the farm, 27).

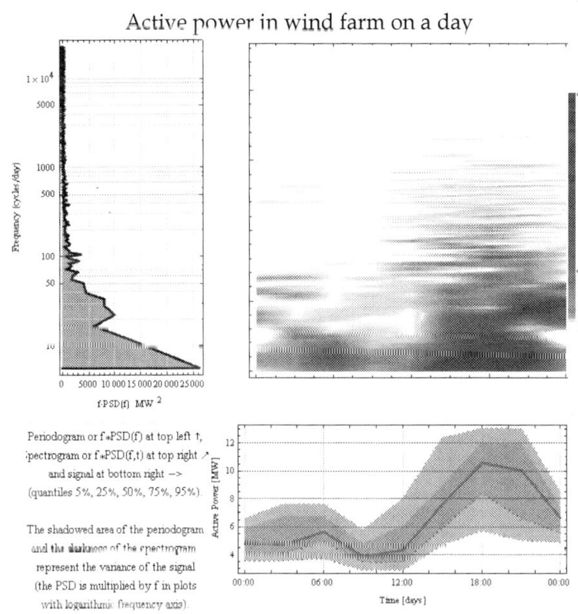

Fig. 16. Spectrogram of the real power [MW] at the substation.

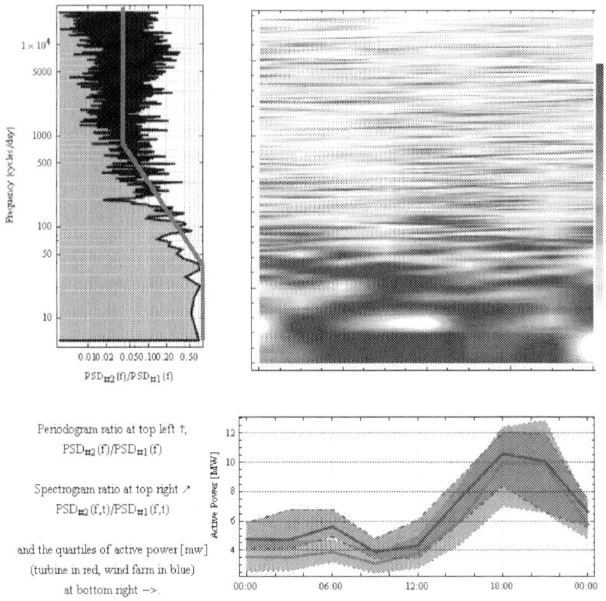

Fig. 17. Squared relative admittance $J^2(f)/N^2$ of the real power of the wind farm relative to the turbine computed as the spectrogram ratio.

	Module of coherence[f]	Appearance
Measured coherence	$\frac{\lvert SFFT_1[f,t] \cdot SFFT_2^*[f,t] \rvert}{\sqrt{PSD_1[f] PSD_2[f]}}$	thin, grey solid line
Fractional model	$\frac{5.58471}{5.58471+0.17906\, f^{0.621445}}$	thick, dotted red line
Power law model	$\frac{1.85021}{f^{0.294854}}$	dot-dashed blue line
Exponential model	$0.187372 + 0.760837\, e^{-0.00584543\, f}$	thin, dashed green line

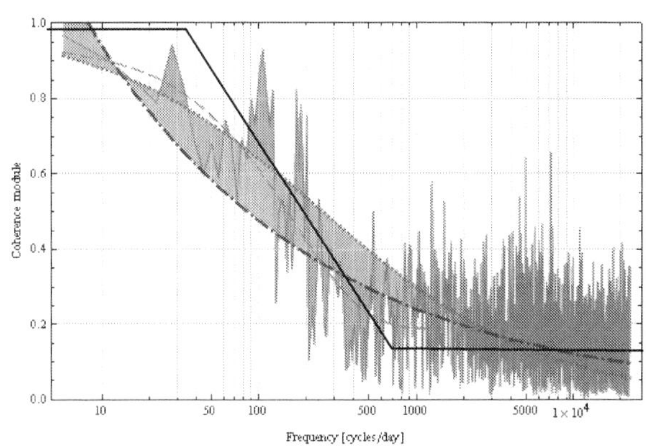

Fig. 18. Coherence models estimated by WINDFREDOM software.

Apart from the Short *FFT* (*SFFT*), the Wigner-Ville distribution (*WVD*) and the S-method (*SM*) have been tested to increase the frequency resolution of the spectrogram. However, the *SFFT* method has been found the most reliable since the amplitudes of the fluctuations are less distorted by the abundant cross-terms present in the power output (Boashash, 2003).

Fig. 15 and Fig. 16 show the spectrogram in the centre of the picture, codified by the scale shown on the right. The plots shown in this subsection have been produced with WINDFREDOM software, which is freely available (Mur-Amada, 2009). The regions with light colours (gray shades in the printed book) indicate that the power has a low content of fluctuations of frequencies corresponding to the vertical axis at the time corresponding to the horizontal axis. The zones with darker colours indicate that fluctuations of the frequency corresponding to the vertical axis have been noticeably observed at the time corresponding to the horizontal axis. For convenience, the median, the quartiles and the 5% and 95% quantiles of the wind speed are also shown in the bottom of the figures. The periodogram is shown on the left and it is computed by averaging the spectrogram.

Both the spectrogram and the periodogram show the auto-spectral density times frequency in Fig. 15 and Fig. 16, because the frequency scale is logarithmic (the derivative of the frequency logarithm is $1/f$). Therefore, the shadowed area of the periodogram or the darkness of the spectrogram is proportional to the variance of the power at each frequency.

Comparing Fig. 15 and Fig. 16, the fluctuations of frequencies higher than 40 cycles/day are relatively smaller in the wind farm than in the turbine. The amount of smoothing at different frequencies is just the squared relative admittance $J^2(f)/N^2$ in Fig. 17. For convenience, $J^2(f)$ has been divided by the number of turbines because $J^2(f)/N^2 \sim 1$ for correlated fluctuations and $J^2(f)/N^2 \sim 1/N$ for uncorrelated fluctuations, (N = 27 is the number of turbines in the wind farm.

The wind farm admittance, corresponding to the periodogram and spectrogram of Fig. 16 divided by Fig. 15 is shown in Fig. 17. The magnitude scale is logarithmic in this plot to remark that the admittance reasonably fits a broken line in a double logarithmic scale.

In this farm, variations quicker than one and three-quarter of a minute (fluctuations of frequency larger than 800 cycles/day) can be considered uncorrelated and fluctuations lasting more than 36 minutes (fluctuations of frequency smaller than 40 cycles/day) can be considered fully correlated. In the intermediate frequency band, the admittance decays as a first order filter, in agreement with the spatial smoothing model.

Fig. 17 shows that the turbine and the wind farm medians (red and blue thick lines in the bottom plot) are similar because slow fluctuations affect both systems alike. The interquartil range (red and blue shadowed areas) is a bit larger in the scaled turbine power with respect to the wind farm. The range has the same magnitude order because the daily variance is primarily due to the correlated fluctuations, since the frequency content of the variance is concentrated in frequencies lower than 40 cycles/day (see grey shadowed area in the periodograms on the left of Fig. 15 and Fig. 16).

In practice, the oscillations measured in the turbine are seen, to some extent, in the substation with some delay or in advance. The coherence $\vec{\gamma}_{\#1,\#2}$ is a complex magnitude with modulus between 0 and 1 and a phase, which represent the delay (positive angles) or the advance (negative angles) of the oscillations of the substation with respect to the turbine. Since the spectrum of a signal is complex, the argument of the coherence $\vec{\gamma}_{i,k}(f)$ is the average phase difference of the fluctuations.

The coherence $\vec{\gamma}_{rc}(f)$ in Fig. 18 indicates the correlation degree and the time pattern of the fluctuations. The modulus is analogous to the correlation coefficient of the spectrum lines from both locations. If the ratio among complex power spectrums is constant (both in modulus and phase), then the coherence is the unity and its argument is the average phase difference. If the complex ratio is random (in modulus or phase), then the coherence is null. The uncertainty of the coherence can be decreased smoothing the plot in Fig. 18. The black broken line is the asymptotic approximation proposed in this chapter and the dashed and dotted lines correspond to other mathematical fits of the coherence.

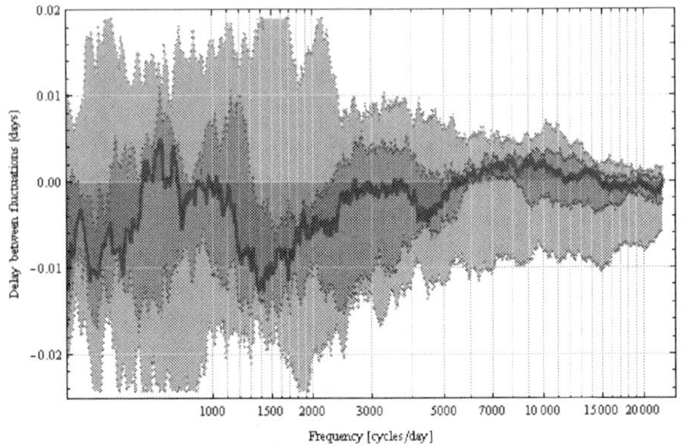

Fig. 19. Time delay quantiles between the fluctuation delays estimated by WINDFREDOM software.

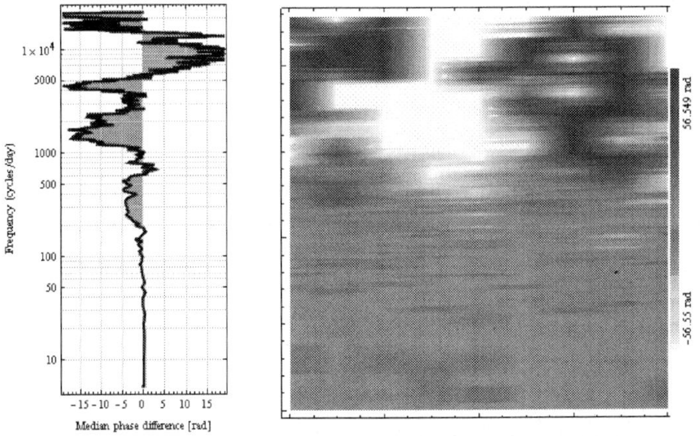

Fig. 20. Estimated phase delay between the power oscillations at the turbine and at the wind farm output. The median value for each frequency *f* is presented on the left and the phase differences of the spectrograms in Fig. 15 and Fig. 16 are presented on the right. A phase unwrapping algorithm has been used to reconstruct the phase from the *SFFT*.

The shadowed area in Fig. 19 indicates the 5%, 25%, 50%, 75% and 95% quantiles of the time delay τ between the oscillations observed at the turbine and the farm output. Fig. 19 shows that the time delay is less than half an hour (0.02 days) the 90% of the time. However, the time delay experiences great variability due to the stochastic nature of turbulence.
Wind direction is not considered in this study because it was steady during the data presented in the chapter. However, the wind direction and the position of the reference turbine inside the farm affect the time delay τ between oscillations. If wind direction changes, the phase difference, $\Delta\varphi = 2\pi f \tau$, can change notably in the transition frequency band, leading to very low coherences in that band. In such cases, data should be divided into series with similar atmospheric properties.
At frequencies lower than 40 cycles/day, the time delays in Fig. 19 implies small phase differences, $\Delta\varphi = 2\pi f \tau$ (colorized in light cyan in Fig. 20), and fluctuations sum almost fully correlated. At frequencies higher than 800 cycles/day, the phase difference $\Delta\varphi = 2\pi f \tau$ usually exceeds several times $\pm 2\pi$ radians (colorized in dark blue or white in Fig. 20), and fluctuations sum almost fully uncorrelated. It should be noticed that the phase difference $\Delta\varphi$ exceeds several revolutions at frequencies higher than 3000 cycles/day and the estimated time delay in Fig. 10 has larger uncertainty (Ghiglia & Pritt, 1998). Thus, the unwrapping phase method could cause the time delay to be smaller at higher frequencies in Fig. 11.
This methodology has been used in (Mur-Amada & Bayod-Rujula, 2010) to compare the wind variations at several weather stations (wind speed behaves more linearly than generated power). The WINDFREDOM software is free and it can be downloaded from www.windygrid.org.

7. Conclusions

This chapter presents some data examples to illustrate a stochastic model that can be used to estimate the smoothing effect of the spatial diversity of the wind across a wind farm on the total generated power. The models developed in this chapter are based in the personal experience gained designing and installing multipurpose data loggers for wind turbines, and wind farms, and analyzing their time series.
Due to turbulence, vibration and control issues, the power injected in the grid has a stochastic nature. There are many specific characteristics that impact notably the power fluctuations between the first tower frequency (usually some tenths of Hertzs) and the grid frequency. The realistic reproduction of power fluctuations needs a comprehensive model of each turbine, which is usually confidential and private. Thus, it is easier to measure the fluctuations in a site and estimate the behaviour in other wind farms.
Variations during the continuous operation of turbines are experimentally characterized for timescales in the range of minutes to fractions of seconds. A stochastic model is derived in the frequency domain to link the overall behaviour of a large number of wind turbines from the operation of a single turbine. Some experimental measurements in the joint time-frequency domain are presented to test the mathematical model of the fluctuations.
The admittance of the wind farm is defined as the ratio of the oscillations from a wind farm to the fluctuations from a single turbine, representative of the operation of the turbines in the farm. The partial cancellation of power fluctuations in a wind farm are estimated from the ratio of the farm fluctuation relative to the fluctuation of one representative turbine.

Provided the Gaussian approximation is accurate enough, the wind farm power variability is fully characterized by its auto spectrum and many interesting properties can be estimated applying the outstanding properties of Gaussian processes (the mean power fluctuation shape during a period, the distribution of power variation in a time period, the most extreme power variation expected during a short period, etc.).

8. References

Abdi A.; Hashemi, H. & Nader-Esfahani, S. (2000). "On the PDF of the Sum of Random Vectors", IEEE Trans. on Communications. Vol. 48, No.1, January 2000, pp 7-12.

Alouini, M.-S.; Abdi, A. & Kaveh, M. (2001). "Sum of Gamma Variates and Performance of Wireless Communication Systems Over Nakagami-Fading Channels", IEEE Trans. On Vehicular Technology, Vol. 50, No. 6, (2001) pp. 1471-1480.

Amarís, H. & Usaola J. (1997). Evaluación en el dominio de la frecuencia de las fluctuaciones de tensión producidas por los generadores eólicos. V Jornadas Hispano-Lusas de Ingeniería Eléctrica. 1997.

Apt, J. (2007) "The spectrum of power from wind turbines", Journal of Power Sources 169 (2007) 369–374

Y. Baghzouz, R. F. Burch et alter (2002) "Time-Varying Harmonics: Part II—Harmonic Summation and Propagation", IEEE Trans. On Power Systems, Vol. 17, No. 1 (January 2002), pp. 279-285.

Bianchi, F. D.; De Battista, H. & Mantz, R. J. (2006). "Wind Turbine Control Systems. Principles, Modelling and Gain Scheduling Design", Springer, 2006.

Bierbooms, W.A.A.M. (2009) "Constrained Stochastic Simulation Of Wind Gusts For Wind Turbine Design", DUWIND Delft University Wind Energy Research Institute, March 2009.

Boashash, B. (2003). "Time Frequency, Signal Analysis and Processing. A comprehensive Reference". Ed. Elsevier, 2003.

Cavers, J.K. (2003). "*Mobile Channel Characteristics*", 2nd ed., Shady Island Press, 2003.

Cidrás, J.; Feijóo, A.E.; González C. C., (2002). "Synchronization of Asynchronous Wind Turbines" IEEE Trans, on Energy Conv., Vol. 17, No 4 (Nov. 2002), pp. 1162-1169

Comech-Moreno, M.P. (2007). "Análisis y ensayo de sistemas eólicos ante huecos de tension", Ph.D. Thesis, Zaragoza University, October 2007 (in Spanish).

Cushman-Roisin, B. (2007). "Environmental Fluid Mechanics", John Wiley & Sons, 2007.

Frandsen, S.; Jørgensen, H.E. & Sørensen, J.D. (2007) "Relevant criteria for testing the quality of turbulence models", 2007 European Wind Energy Conference and Exhibition, Milan (IT), 7-10 May 2007. pp. 128-132.

Gardner, W. A. (1994) "Cyclostationarity in Communications and Signal Processing", IEEE press, 1994.

Gardnera, W. A.; Napolitano, A. & Paurac, L. (2006) "Cyclostationarity: Half a century of research", Signal Processing 86 (April 2006), pp. 639–697.

Ghiglia, D.C. & Pritt, M.D. (1998). "Two-Dimensional Phase Unwrapping: Theory, Algorithms, and Software", John Whiley & Sons, 1998.

Hall, P.; & Heyde. C. C. (1980). Martingale Limit Theory and Its Application. New York: Academic Press (1980).

Kaimal, J.C. (1978). "Horizontal Velocity Spectra in an Unstable Surface Layer" Journal of the Atmospheric Sciences, Vol. 35, Issue 1 (January 1978), pp. 18–24.

Karaki, S. H. ; Salim B. A. & Chedid R. B. (2002). "Probabilistic Model of a Two-Site Wind Energy Conversion System", IEEE Transactions On Energy Conversion, Vol. 17, No. 4, December 2002.

Kundur, P. P.; Balu, N. J.; Lauby, M. G. (1994). "Power System Stability and Control", McGraw-Hill, 1994.

Li, P.; Banakar, H.; Keung, P. K.; Far H.G. & Ooi B.T. (2007). "Macromodel of Spatial Smoothing in Wind Farms", IEEE Trans, on Energy Conv., Vol. 22, No 1 (March. 2007), pp 119-128.

Martins, A.; Costa, P.C. & Carvalho, A. S. (2006). "Coherence And Wakes In Wind Models For Electromechanical And Power Systems Standard Simulations", European Wind Energy Conferences (EWEC 2006), February (2006), Athens.

Mur-Amada, J. (2009) "Wind Power Variability in the Grid", PhD. Thesis, Zaragoza University, October 2009. Available at www.windygrid.org

Mur-Amada, J. & Comech-Moreno, M.P. (2006). "Reactive Power Injection Strategies for Wind Energy Regarding its Statistical Nature", Sixth International Workshop on Large-Scale Integration of Wind Power and Transmission Networks for Offshore Wind Farm. Delft, October 2006.

Mur-Amada, J. & Bayod-Rújula, A.A. (2007). "Characterization of Spectral Density of Wind Farm Power Output", *9th Conference on Electrical Power Quality and Utilisation (EPQU'2007)*, Barcelona, October 2007.

Mur-Amada, J. & Bayod-Rújula, A.A. (2010). "Variability of Wind and Wind Power", *Wind Power*, Intech, Croatia, 2010. Available at: www.sciyo.com.

Norgaard, P. & Holttinen, H. (2004). "A Multi-turbine Power Curve Approach", in Proc. 2004 Nordic Wind Power Conference (NWPC 2002), Gothenberg, March 2004.

Press, W. H.; Teukolsky, S. A.; Vetterling, W. T. & Flannery, B. P. (2007). "Numerical Recipes The Art of Scientific Computing", 3rd edition, Cambridge University Press, 2007.

Sanz M.; Llombart A.; Bayod A. A. & Mur, J. (2000) "Power quality measurements and analysis for wind turbines", *IEEE Instrumentation and Measurement Technical Conference 2000*, pp. 1167-1172. May 2000, Baltimore.

Saranyasoontorn, K.; Manuel, L. & Veers, P. S. "A Comparison of Standard Coherence Models form Inflow Turbulence With Estimates from Field Measurements", Journal of Solar Energy Engineering, Vol. 126 (2004), Issue 4, pp. 1069-1082

Schlez, W. & Infield, D. (1998). "Horizontal, two point coherence for separations greater than the measurement height", *Boundary-Layer Meteorology* 87 (1998), 459-480.

Schwab, M.; Noll, P. & Sikora, T. (2006). "Noise robust relative transfer function estimation", XIV European Signal Processing Conference, September 4 - 8, 2006, Florence, Italy.

Soens, J. (2005). "Impact Of Wind Energy In A Future Power Grid", Ph.D. Dissertation, Katholieke Universiteit Leuven, December 2005.

Sorensen, P.; Hansen, A. D. & Rosas C. (2002). "Wind models for simulation of power fluctuations from wind farms", Journal of Wind Engineering and Ind. Aerodynamics 90 (2002), pp. 1381-1402

Sørensen, P.; Cutululis, N. A.; Vigueras-Rodríguez, A; Madsen, H.; Pinson, P; Jensen, L. E.; Hjerrild, J. & Donovan M., (2008) "Modelling of Power Fluctuations from Large Offshore Wind Farms", *Wind Energy*,Volume 11, Issue 1, pages 29-43, January/February 2008.

Stefopoulos, G.; Meliopoulos A. P.& Cokkinides G. J. (2005), "Advanced Probabilistic Power Flow Methodology", 15th PSCC, Liege, 22-26 August 2005

Su, C-L. (2005) "Probabilistic Load-Flow Computation Using Point Estimate Method", IEEE Trans. Power Systems, Vol. 20, No. 4, November 2005, pp. 1843-1851.

Tentzerakis, S. T. & Papathanassiou S. A. (2007), "An Investigation of the Harmonic Emissions of Wind Turbines", IEEE Trans, on Energy Conv., Vol. 22, No 1, March. 2007, pp 150-158.

Thiringer, T.; Petru, T.; & Lundberg, S. (2004) "Flicker Contribution From Wind Turbine Installations" IEEE Trans, on Energy Conv., Vol. 19, No 1, March 2004, pp 157-163.

Vilar Moreno, C. (2003). "Voltage fluctuation due to constant speed wind generators" Ph.D. Thesis, Carlos III University, Leganés, Spain, 2003.

Wangdee, W. & Billinton R. (2006). "Considering Load-Carrying Capability and Wind Speed Correlation of WECS in Generation Adequacy Assessment", IEEE Trans, on Energy Conv., Vol. 21, No 3, September 2006, pp. 734-741.

Welfonder, E.; Neifer R. & Spaimer, M. (1997) "Development And Experimental Identification Of Dynamic Models For Wind Turbines", Control Eng. Practice, Vol. 5, No. 1 (January 2007), pp. 63-73.

Part 4

Input into Power System Networks

7

Distance Protections in the Power System Lines with Connected Wind Farms

Adrian Halinka and Michał Szewczyk
Silesian University of Technology
Poland

1. Introduction

In recent years there has been an intensive effort to increase the participation of renewable sources of electricity in the fuel and energy balance of many countries. In particular, this relates to the power of wind farms (*WF*) attached to the power system at both the distribution network (the level of MV and 110 kV) and the HV transmission network (220 kV and 400 kV)[1]. The number and the level of power (from a dozen to about 100 MW) of wind farms attached to the power system are growing steadily, increasing the participation and the role of such sources in the overall energy balance. Incorporating renewable energy sources into the power system entails a number of new challenges for the power system protections in that it will have an impact on distance protections which use the impedance criteria as the basis for decision-making. The prevalence of distance protections in the distribution networks of 110 kV and transmission networks necessitates an analysis of their functioning in the new conditions. This study will be considering selected factors which influence the proper functioning of distance protections in the distribution networks with the wind farms connected to the power system.

2. Interaction of dispersed power generation sources (DPGS) with the power grid

There are two main elements determining the character of work of the so-called dispersed generation objects with the power grid. They are the type of the generator and the way of connection.

In the case of using asynchronous generators, only parallel "cooperation" with the power system is possible. This is due to the fact that reactive power is taken from the system for magnetization. When the synchronous generator is used or the generator is connected by the power converter, both parallel or autonomous (in the power island) work is possible.

The level of generating power and the quality of energy have to be taken into consideration when dispersed power sources are to be connected to the distribution network. In regard to wind farms, it should be emphasized that they are mainly connected to the HV distribution

---

[1] The way of connection and power grid configuration differs in many countries. Sample configurations are taken from the Polish Power Grid but can be easily adapted to the specific conditions in the particular countries.

network for the reason of their relatively high generating power and not the best quality of energy. This connection is usually made by the HV to MV transformer. It couples an internal wind farm electrical network (on the MV level) with the HV distribution network. The internal wind farm network consists of cable MV lines working in the trunk configuration connecting individual wind turbines with the coupling HV/MV transformer. Fig. 1 shows a sample structure of the internal wind farm network.

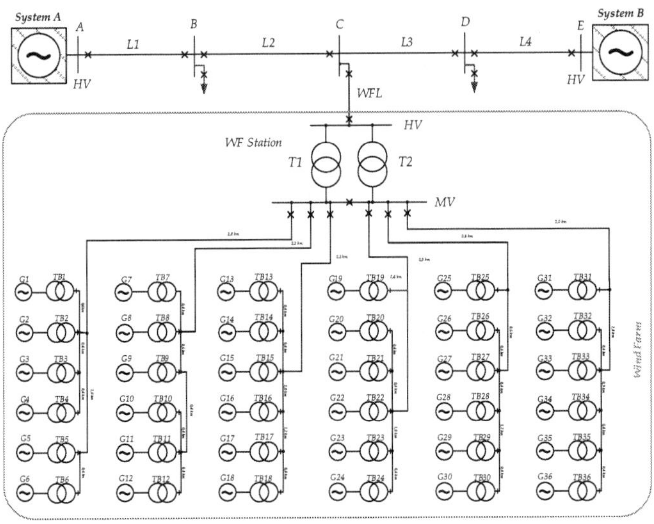

Fig. 1. Sample structure of internal electrical network of the 72 MW wind farm connected to the HV distribution network

There are different ways of connecting wind farms to the HV network depending, among other things, on the power level of a wind farm, distance to the HV substation and the number of wind farms connected to the sequencing lines. One can distinguish the following characteristic types of connections of wind farms to the transmission network:

- Connection in the three-terminal scheme (Fig. 2a). For this form of connection the lowest investment costs can be achieved. On the other hand, this form of connection causes several serious technical problems, especially for the power system automation. They are related to the proper faults detection and faults elimination in the surroundings of the wind farm connection point. Currently, this is not the preferred and recommended type of connection. Usually, the electrical power of such a wind farm does not exceed a dozen or so MW.
- Connection to the HV busbars of the existing substation in the series of lines (Fig. 2b). This is the most popular solution. The level of connected wind farms is typically in the range of 5 to 80 MW.
- Connection by the cut of the line (Fig. 3.). This entails building a new substation. If the farm is connected in the vicinity of an existing line, a separate wind farm feeder line is superfluous. Only cut ends of the line have to be guided to the new wind farm power substation. This substation can be made in the H configuration or the more complex 2

circuit-breaker (2CB) configuration (Fig. 3b). The topology of the substation depends on the number of the target wind farms connected to such a substation.

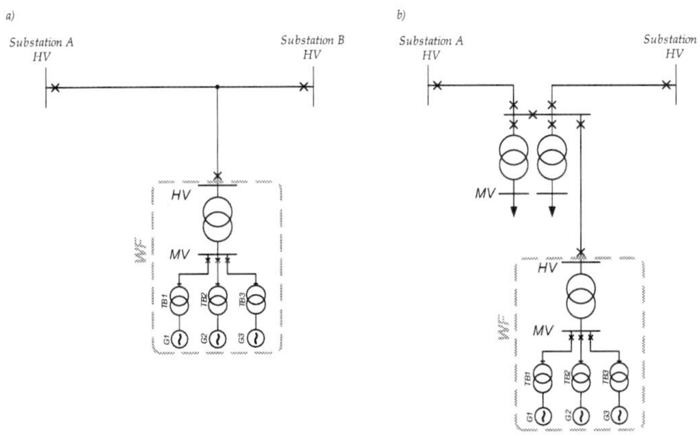

Fig. 2. Types of the wind farm connection to HV network: a) three terminal-line , b) connection to the busbars of existing HV/MV substation

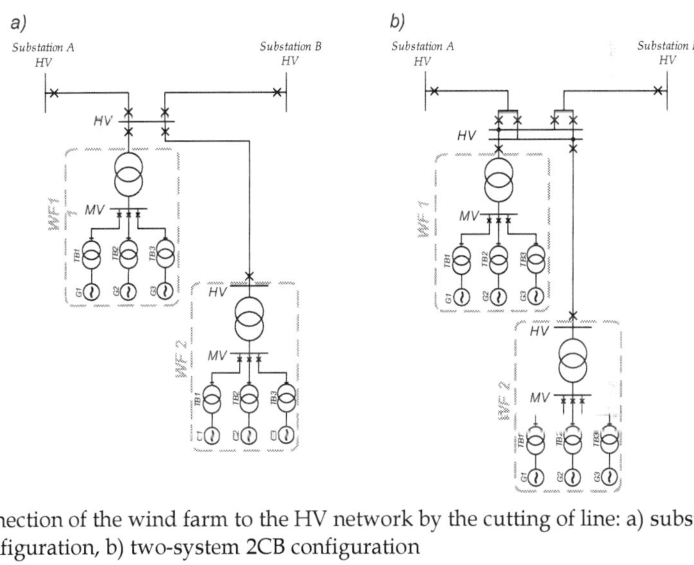

Fig. 3. Connection of the wind farm to the HV network by the cutting of line: a) substation in the H4 configuration, b) two-system 2CB configuration

- Connection to the HV switchgear of the EHV/HV substation bound to the transmission network. In this case one of the existing HV line bays (Fig. 4a) or the separate transformer (Fig. 4b) can be used. This form of connection is possible for wind farms of high level generating powers (exceeding 100 MW). The influence of such a connection on the proper functioning of the power protections is the lowest one.

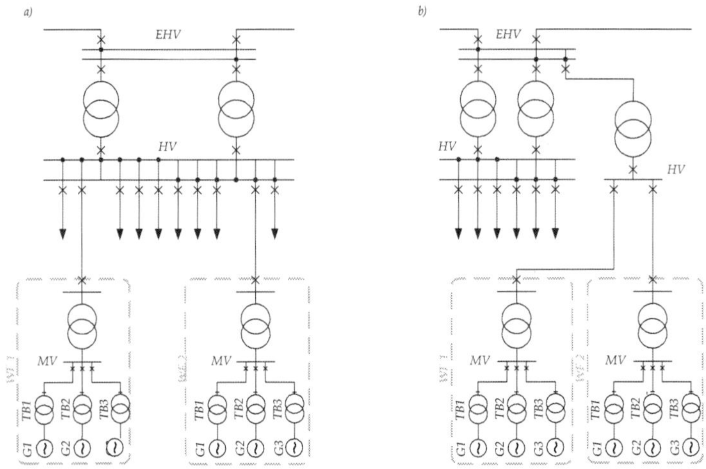

Fig. 4. Wind farm connection to the power system: a) by the existing switching bay of the EHV/HV substation, b) by the HV busbars of the separate EHV/HV transformer

- Connection of the wind farm by the high voltage AC/DC link (Fig. 5). This form is most commonly used for wind farms located on the sea and for different reasons cannot work synchronously with the electrical power system. Using a direct current link is useful for the control of operating conditions of the wind farm, however at the price of higher investments costs.

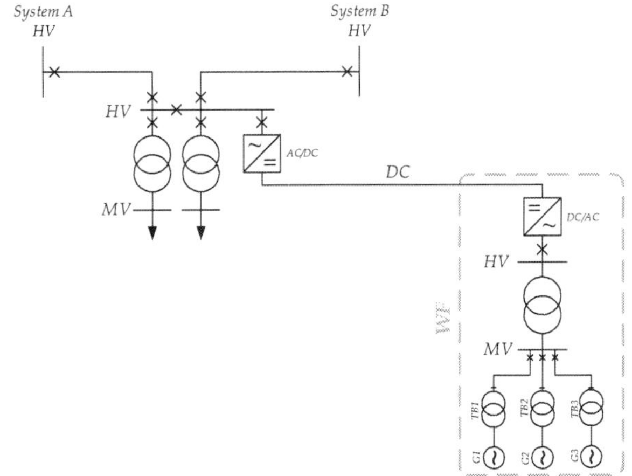

Fig. 5. Connection of the wind farm by the AC/DC link

Due to the limited number of system EHV/HV substations and the relatively high distances between substations and wind farms, most of them are connected to the existing or newly built HV/MV substations inside the HV line series.

3. Technical requirements for the dispersed power sources connected to the distribution network

Basic requirements for dispersed power sources are stipulated by a number of directives and instructions provided by the power system network operator. They contain a wide spectrum of technical conditions which must be met when such objects are connected to the distribution network. From the point of view of the power system automation, these requirements are mainly concerned with the possibilities of the power level and voltage regulation. Additionally, the behaviour of a wind farm during faults in the network and the functioning of power protection automation have to be determined. Wind farms connected to the HV distribution network should be equipped with the remote control, regulation and monitoring systems which enable following operation modes:

- operation without limitations (depending on the weather conditions),
- operation with an assumed *a priori* power factor and limited power generation,
- intervention operation during emergences and faults in the power system (type of intervention is defined by the operator of the distribution network),
- voltage regulator at the connection point,
- participation in the frequency regulation (this type of work is suitable for wind farms of the generating power greater than 50 MW).

During faults in HV network, when significant changes (dips) of voltage occur, wind farm cannot loose the capability for reactive power regulation and should actively work towards sustaining the voltage level in the network. It also should maintain continuous operation in the case of faults in the distribution network which cause voltage dips at the wind farm connection point, of the times over the borderline shown in Fig. 6.

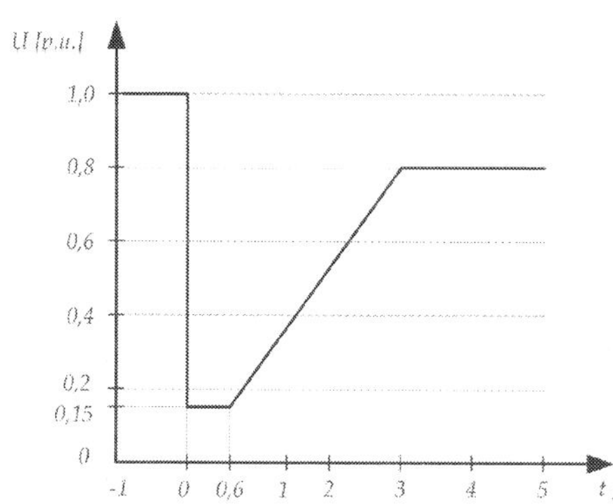

Fig. 6. Borderline of voltage level conditioning continuous wind farm operation during faults in the distribution network

4. Dispersed power generation sources in fault conditions

The behaviour of a power system in dynamic fault states is much more complicated for the reason of the presence of dispersed power sources than when only the conventional ones are in existence. This is a direct consequence of such factors as the technical construction of driving units, different types of generators, the method of connection to the distribution network, regulators and control units, the presence of fault ride-through function as well as a wide range of the generating power determined by e.g. the weather conditions.

Taking the level of fault current as the division criteria, the following classification of dispersed power sources can be suggested:

- sources generating a constant fault current on a much higher level than the nominal current (mainly sources with synchronous generators),
- sources generating a constant fault current close to the nominal current (units with DFIG generators or units connected by the power converters with the fault ride-through function),
- sources not designed for operation in faulty conditions (sources with asynchronous generators or units with power converters without the fault ride-through function).

Sources with synchronous generators are capable of generating a constant fault current of higher level than the nominal one. This ability is connected with the excitation unit which is employed and with the voltage regulator. Synchronous generators with an electromechanical excitation unit are capable of holding up a three-phase fault current of the level of three times or higher than the nominal current for a few seconds. For the electronic (static) excitation units, in the case of a close three-phase fault, it is dropping to zero after the disappearance of transients. This is due to the little value of voltage on the output of the generator during a close three-phase fault.

For asynchronous generators, the course of a three-phase current on its outputs is only limited by the fault impedance. The fault current drops to zero in about (0,2 ÷ 0,3) s. The maximum impulse current is close to the inrush current during the motor start-up of the generator (Lubośny, 2003). The value of such a current for typical machines is five times higher than the nominal current. This property makes it possible to limit the influence of such sources only on the initial value of the fault current and value of the impulse current.

The construction and parameters of the power converters in the power output circuit determine the level of fault current for such dispersed power sources. Depending on the construction, they generate a constant fault current on the level of its nominal current or are immediately cut off from the distribution network after a detection of a fault. If the latter is the case, only a current impulse is generated just after the beginning of a fault.

A common characteristic of dispersed sources cooperating with the power system is the fact that they can achieve local stability. Some of the construction features (power converters) and regulatory capabilities (reactive power, frequency regulation) make the dispersed power generation sources units highly capable of maintaining the stability in the local network area during the faulty conditions (Lubośny, 2003).

Dynamic states analyses must take into consideration the fact that present wind turbines are characterized by much higher resistance to faults (voltage dips) to be found in the power system than the conventional power sources based on the synchronous generators. A very important and useful feature of some wind turbines equipped with power converters, is the fact that they can operate in a higher frequency range (43 ÷ 57 Hz) than in conventional sources (47 ÷ 53 Hz) (Ungrad et al., 1995).

Dispersed generation may have a positive influence on the stability of the local network structures: *dispersed source – distribution network* during the faults. Whether or not it can be well exploited, depends on the proper functioning of the power system protection automation dedicated to the distribution network and dispersed power generation sources.

5. Influence of connecting dispersed power generating sources to the distribution network on the proper functioning of power system protections

In the Polish power system most of generating power plants (the so-called system power plants) are connected to the HV and EHV (220 kV and 400 kV) transmission networks. Next, HV networks are usually treated as distribution networks powered by the HV transmission networks. This results in the lack of adaptation of the power system protection automation in the distribution network to the presence of power generating sources on those (MV and HV) voltage levels.

Even more frequently, using of the DPGS, mainly wind farms, is the source of potential problems with the proper functioning of power protection automation. The basic functions vulnerable to the improper functioning in such conditions are:

- primary protection functions of lines,
- earth-fault protection functions of lines,
- restitution automation, especially auto-reclosing function,
- overload functions of lines due the application of high temperature low sag conductors and the thermal line rating,
- functions controlling an undesirable transition to the power island with the local power generation sources.

The subsequent part of this paper will focus only on the influence of the presence of the wind farms on the correctness of action of impedance criteria in distance protections.

5.1 Selected aspects of an incorrect action of the distance protections in HV lines

Distance protection provides short-circuit protection of universal application. It constitutes a basis for network protection in transmission systems and meshed distribution systems. Its mode of operation is based upon the measurement and evaluation of the short-circuit impedance, which in the typical case is proportional to the distance to the fault. They rarely use pilot lines in the 110 kV distribution network for exchange of data between the endings of lines. For the primary protection function, comparative criteria are also used. They take advantage of currents and/or phases comparisons and use of pilot communication lines. However, they are usually used in the short-length lines (Ungrad et al., 1995).

The presence of the DPGS (wind farms) in the HV distribution network will affect the impedance criteria especially due to the factors listed below:

- highly changeable value of the fault current from a wind farm. For wind farms equipped with power converters, taking its reaction time for a fault, the fault current is limited by them to the value close to the nominal current after typically not more then 50 ms. So the impact of that component on the total fault current evaluated in the location of protection is relatively low.
- intermediate in-feed effect at the wind farm connection point. For protection realizing distance principles on a series of lines, this causes an incorrect fault localization both in the primary and the back-up zones,
- high dynamic changes of the wind farm generating power. Those influence the more frequent and significant fluctuations of the power flow in the distribution network. They are not only limited to the value of the load currents but also to changes of their directions. In many cases a load of high values must be transmitted. Thus, it is necessary to use wires of higher diameter or to apply high temperature low sag

conductors or thermal line rating schemes (dynamically adjusting the maximum load to the seasons or the existing weather conditions). Operating and load area characteristics may overlap in these cases.

Setting distance protections for power lines

In the case of distance protections, a three-grading plan (Fig. 7) is frequently used. Additionally, there are also start-up characteristic and the optional reverse zone which reach the busbars.

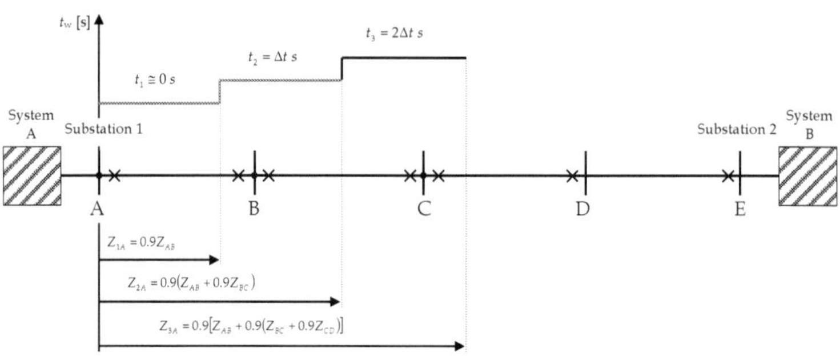

Fig. 7. Three-grading plan of distance protection on series of lines

The following principles can be used when the digital protection terminal is located in the substation A (Fig. 7) (Ziegler, 1999):

- impedance reach of the first zone is set to 90 % of the A-B line-length

$$Z_{1A} = \left|0.9\underline{Z}_{AB}\right| \tag{1}$$

tripping time t_1=0 s;

- impedance reach of the second zone cannot exceed the impedance reach of the first zone of protection located in the substation B

$$Z_{2A} = \left|0.9\left(\underline{Z}_{AB} + 0.9\underline{Z}_{BC}\right)\right| \tag{2}$$

tripping time should be one step higher than the first one t_2=Δt s from the range of (0.3÷0.5) s. Typically for the digital protections and fast switches, a delay of 0.3 s is taken;

- impedance reach of the third zone is maximum 90% of the second zone of the shortest line outgoing from the subsubstation B:

$$Z_{3A} = \left|0 \quad \left[\underline{Z}_{AB} + \quad \left(\underline{Z}_{BC} + \quad \underline{Z}_{CD}\right)\right]9\right| \qquad 0 \tag{3}$$

For the selectivity condition, tripping time for this zone cannot by shorter than t_3=2Δt s.

Improper fault elimination due to the low fault current value

As mentioned before, when the fault current flowing from the DPGS is close to the nominal current, in most of cases overcurrent and distance criteria are difficult or even impossible to apply for the proper fault elimination (Pradhan & Geza, 2007). Figure 8 presents sample courses of the rms value of voltage U, current I, active and reactive power (P and Q) when there are voltage dips caused by faults in the network. The recordings are from a wind turbine equipped with a 2 MW generator with a fault ride-through function (Datasheet, Vestas). This function permits wind farm operation during voltage dips, which is generally required for wind farms connected to the HV networks.

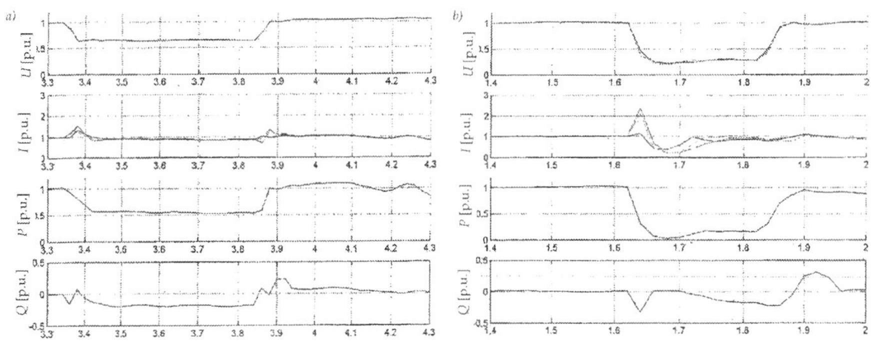

Fig. 8. Courses of electric quantities for Vestas V80 wind turbine of 2 MW: a) voltage dip to 0.6 U_N, b) voltage dip to 0.15 U_N (Datasheet, Vestas)

Analyzing the course of the current presented in Fig. 8, it can be observed that it is close to the nominal value and in fact independent a of voltage dip. Basing on the technical data it is possible to approximate t_1 time, when the steady-state current will be close to the nominal value (Fig. 9).

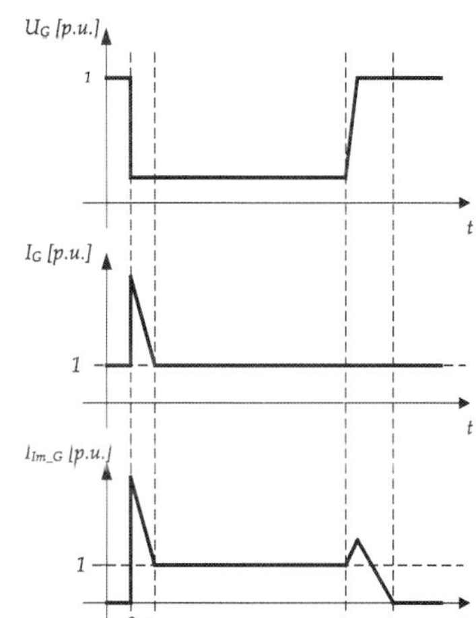

Fig. 9. Linear approximation of current and voltage values for the wind turbine with DFIG generator during voltage dips: U_G - voltage on generator outputs, I_G - current on generator outputs, I_{Im_G} - generator reactive current, $t_1 \approx 50$ ms, t_3-$t_2 \approx 100$ ms

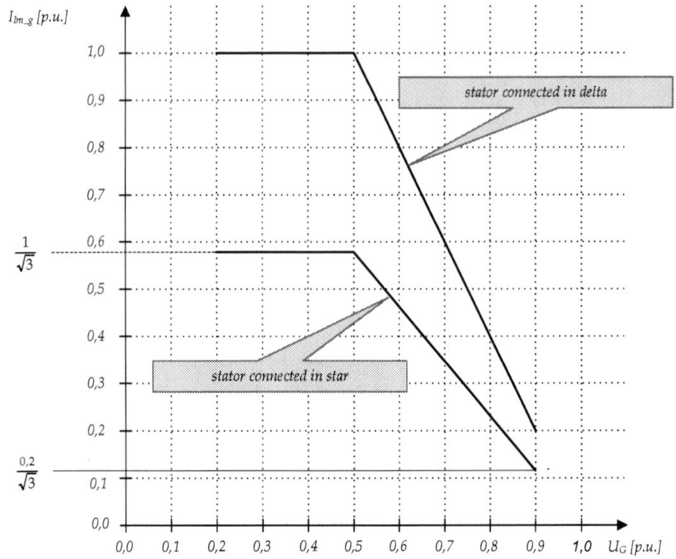

Fig. 10. Course of the wind turbine reactive current

The negative influence of the low value steady current from the wind farm is cumulating especially when the distribution network is operating in the open configuration (Fig. 11).

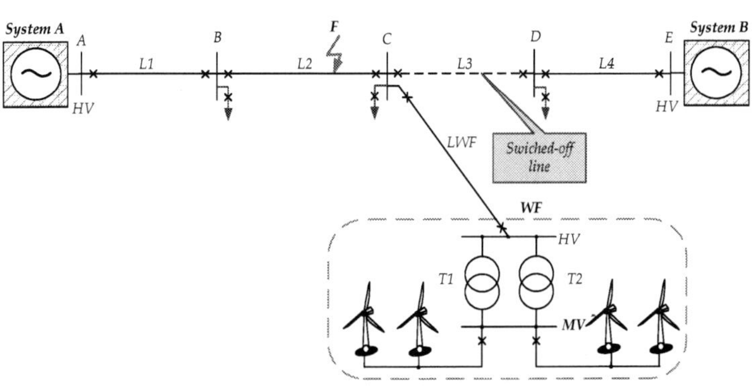

Fig. 11. Wind farm in the distribution network operating in the open configuration

The selected wind turbine is the one most frequently used in the Polish power grid. The impulse current at the beginning of the fault is reduced to the value of the nominal current after 50 ms. Additionally, the current has the capacitance character and is only dependent on the stator star/delta connection. This current has the nominal value for delta connection (high rotation speed of turbine) and nominal value divided by $\sqrt{3}$ for the star connection as presented in Fig. 9.

Reaction of protection automation systems in this configuration can be estimated comparing the fault current to the pick-up currents of protections. For a three-phase fault at point F (Fig. 11) the steady fault current flowing through the wind farm cannot exceed the nominal current of the line. The steady fault current of the single wind turbine of P_N=2 MW (S_N=2.04 MW) is $I_k = I_{NG} = 10.7$ A at the HV side (delta stator connection). However initial fault current I_k'' is 3,3 times higher than the nominal current ($I_k'' = 35.31$ A).It must be emphasized that the number of working wind turbines at the moment of a fault is not predictable. This of course depends on weather conditions or the network operator's requirements. All these influence a variable fault current flowing from a wind farm. In many cases there is a starting function of the distance protection in the form of a start-up current at the level of 20% of the nominal current of the protected line. Taking 600 A as the typical line nominal current, even several wind turbines working simultaneously are not able to exceed the pick-up value both in the initial and the steady state fault conditions. When the impedance function is used for the pick-up of the distance protection, the occurrence of high inaccuracy and fluctuations of measuring impedance parameters are expected, especially in the transient states from the initial to steady fault conditions.

The following considerations will present a potential vulnerability of the power system distribution networks to the improper (missing) operation of power line protections with connected wind farms. In such situations, when there is a low fault current flow from a wind farm, even using the alternative comparison criteria will not result in the improvement of its operation. It is because of the pick-up value which is generally set at (1,2 ÷ 1,5) I_N.

To minimize the negative consequences of functioning of power system protection automation in HV network operating in an open configuration with connected wind farms, the following instructions should be taken:

- limiting the generated power and/or turning off the wind farm in the case of a radial connection of the wind farm with the power system. In this case, as a result of planned or fault switch-offs, low fault WF current occurs,
- applying distance protection terminals equipped with the weak end infeed logic on all of the series of HV lines, on which the wind farm is connected. The consequences are building up the fast teletransmission network and relatively high investment costs,
- using banks of settings, configuring adaptive distance protection for variant operation of the network structure causing different fault current flows. When the HV distribution network is operating in a close configuration, the fault currents considerably exceed the nominal currents of power network elements. In the radial configuration, the fault current which flows from the local power source will be under the nominal value.

Selected factors influencing improper fault location of the distance protections of lines

In the case of modifying the network structure by inserting additional power sources, i.e. wind farms, the intermediate in-feeds occur. This effect is the source of impedance paths measurement errors, especially when a wind farm is connected in a three-terminal configuration. Figure 12a shows the network structure and Fig. 12b a short-circuit equivalent scheme for three-phase faults on the M-F segment. Without considering the measuring transformers, voltage $\underline{U}_p$ in the station A is:

$$\underline{U}_p - \underline{Z}_{AM}\underline{I}_A + \underline{Z}_{MF}\underline{I}_Z - \underline{Z}_{AM}\underline{I}_A + \underline{Z}_{MF}\left(\underline{I}_A + \underline{I}_{WF}\right) \qquad (4)$$

On the other hand current $\underline{I}_p$ measured by the protection in the initial time of fault is the fault current $\underline{I}_A$ flowing in the segment A-M. Thus the evaluated impedance is:

$$\underline{Z}_p = \frac{\underline{U}_p}{\underline{I}_p} = \frac{\underline{Z}_{AM}\underline{I}_A + \underline{Z}_{MF}\left(\underline{I}_A + \underline{I}_{WF}\right)}{\underline{I}_A} = \underline{Z}_{AM} + \underline{Z}_{MF}\left(1 + \frac{\underline{I}_{WF}}{\underline{I}_A}\right) = \underline{Z}_{AM} + \underline{Z}_{MF}\underline{k}_{if} \quad (5)$$

where:

$\underline{U}_p$ - positive sequence voltage component on the primary side of voltage transformers at point A,

$\underline{I}_p$ - positive sequence current component on the primary side of current transformers at point A,

$\underline{I}_A$ - fault current flowing from system A,

$\underline{I}_{WF}$ - fault current flowing from WF,

$\underline{Z}_{AM}$ - impedance of the AM segment,

$\underline{Z}_{MF}$ - impedance of the MF segment,

k_{if} - intermediate in-feed factor.

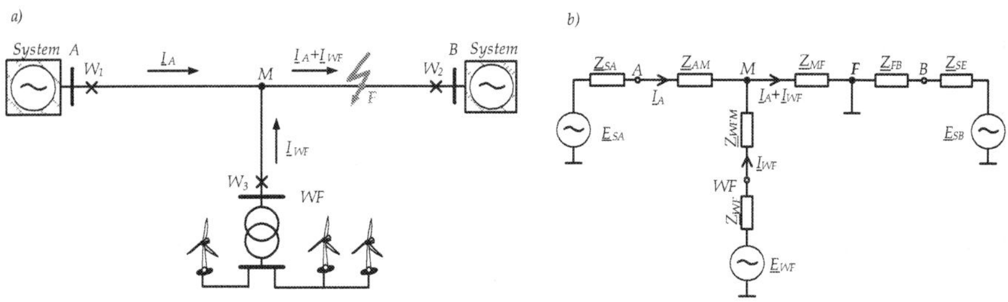

Fig. 12. Teed feeders configuration a) general scheme, b) equivalent short-circuit scheme.

It is evident that estimated from (5) impedance is influenced by error ΔZ:

$$\underline{\Delta Z} = \underline{Z}_{MF}\frac{\underline{I}_{WF}}{\underline{I}_A} \quad (6)$$

The error level is dependent on the quotient of fault current $\underline{I}_Z$ from system A and power source WF (wind farm). Next the error is always positive so the impedance reaches of the operating characteristics are shorter. Evaluating the error level from the impedance of the equivalent short-circuit:

$$\underline{\Delta Z} = \underline{Z}_{MF}\frac{\underline{Z}_{SA} + \underline{Z}_{AM}}{\underline{Z}_{WF} + \underline{Z}_{WFM}} \quad (7)$$

Equation (7) shows the significant impact on the error level of short-circuit powers (impedances of power sources), location of faults ($\underline{Z}_{AM}$, $\underline{Z}_{FWM}$) and types of faults.

Minimizing possible errors in the evaluation of impedance can be achieved by modifying the reaches of operating characteristics covering the WF location point. Thus the reaches of the second and the third zone of protection located at point A (Fig. 7) are:

$$Z_{2A} = \left|0.9\left(\underline{Z}_{AB} + 0.9\underline{Z}_{BC}\underline{k}_{if}\right)\right| = \left|0.9\left[\underline{Z}_{AB} + 0.9\underline{Z}_{BC}\left(1 + \frac{\underline{I}_{WF}}{\underline{I}_A}\right)\right]\right| \quad (8)$$

$$Z_{3A} = \left|0.9\left[Z_{AB} + 0.9\left(Z_{BC} + 0.9Z_{CD}\right)\underline{k}_{if}\right]\right| = \left|0.9\left[Z_{AB} + 0.9\left(Z_{BC} + 0.9Z_{CD}\right)\left(1 + \frac{\underline{I}_{WF}}{\underline{I}_A}\right)\right]\right| \quad (9)$$

It is also necessary to modify of the first zone, i.e.:

$$Z_{1A} = \left|0.9\underline{Z}_{AB}\underline{k}_{if}\right| = \left|0.9\underline{Z}_{AB}\left(1 + \frac{\underline{I}_{WF}}{\underline{I}_A}\right)\right| \quad (10)$$

This error correction is successful if the error level described by equations (6) and (7) is constant. But for wind farms this is a functional relation. The arguments of the function are, among others, the impedance of WF $\underline{Z}_{WF}$ and a fault current $\underline{I}_{WF}$. These parameters are dependent on the number of operating wind turbines, distance from the ends of the line to the WF connection point (point M in Fig. 12a), fault location and the time elapsed from the beginning of a fault (including initial or steady fault current of WF).

As mentioned before, the three-terminal line connection of the WF in faulty conditions causes shortening of reaches of all operating impedance characteristics in the direction to the line. This concerns both protections located in substation A and WF. For the reason of reaching reduction level, it can lead to:

- extended time of fault elimination, e.g. fault elimination will be done with the time of the second zone instead of the first one,
- improper fault elimination during the auto-reclosure cycles. This can occurs when during the intermediate in-feed the reaches of the first extended zones overcome shortening and will not reach full length of the line. Then what cannot be reached is simultaneously cutting-off the fault current and the pick-up of auto-reclosure automation on all the line ends.

In Polish HV distribution networks the back-up protection is usually realized by the second and third zones of distance protections located on the adjacent lines. With the presence of the WF (Fig. 13), this back-up protection can be ineffective.

As an example, in connecting WF to substation C operating in a series of lines A-E what should be expected is the miscalculation of impedances in the case of intermediate in-feed in substation C from the direction of WF. The protection of line L2 located in substation B, when the fault occurs at point F on the line L3, "sees" the impedance vector in its second or third zone. The error can be obtained from the equation:

$$\underline{Z}_{pB} = \frac{\underline{U}_{pB}}{\underline{I}_{pB}} = \frac{\underline{I}_{L2}\underline{Z}_{BC} + \left(\underline{I}_{L2} + \underline{I}_{WF}\right)\underline{Z}_{CF}}{\underline{I}_{L2}} = \underline{Z}_{BC} + \underline{Z}_{CF} + \Delta\underline{Z}_{pB} \quad (11)$$

where:

$\underline{U}_{pB}$ - positive sequence voltage on the primary side of voltage transformers at point B,

$\underline{I}_{pB}$ - positive sequence current on the primary side of current transformers at point B,

$\underline{I}_{L2}$ - fault current flowing by the line L2 from system A,

$\underline{I}_{WF}$ - fault current from WF,

$\underline{Z}_{BC}$ - line L2 impedance,
$\underline{Z}_{CF}$ - impedance of segment CF of the line L3
and the error ΔZ_{pB} is defined as:

$$\Delta \underline{Z}_{pB} = \underline{Z}_{CF}\left(\frac{\underline{I}_{WF}}{\underline{I}_{L2}}\right). \tag{12}$$

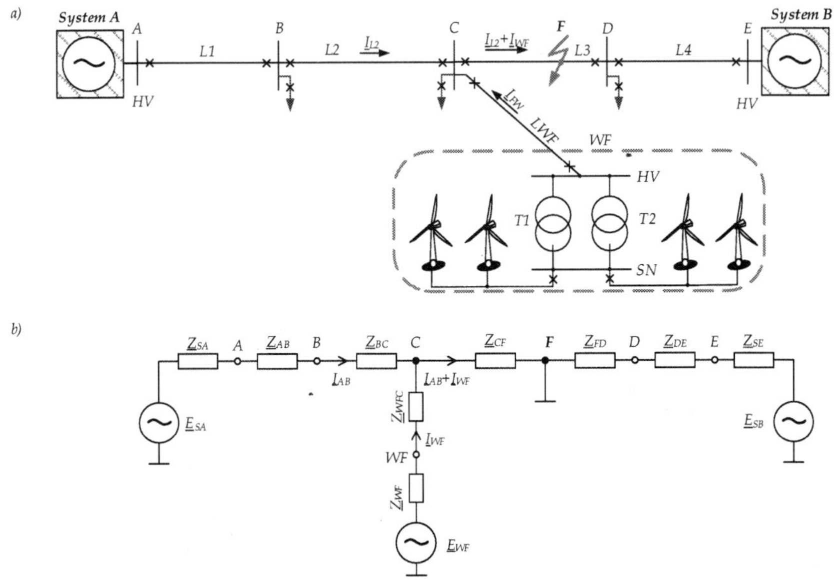

Fig. 13. Currents flow after the WF connection to substation C: a) general scheme, b) simplified equivalent short-circuit scheme

It must be emphasized that, as before, also the impedance reaches of second and third zones of LWF protection located in substation WF are reduced due to the intermediate in-feed. Due to the importance of the back-up protection, it is essential to do the verification of the proper functioning (including the selectivity) of the second and third zones of adjacent lines with wind farm connected. However, due to the functional dynamic relations, which cause the miscalculations of the impedance components, preserving the proper functioning of the distance criteria is hard and requires strong teleinformatic structure and adaptive decision-making systems (Halinka et al., 2006).

Overlapping of the operating and admitted load characteristics

The number of connected wind farms has triggered an increase of power transferred by the HV lines. As far as the functioning of distance protection is concerned, this leads to the increase of the admitted load of HV lines and brings closer the operating and admitted load characteristics. In the case of non-modified settings of distance protections this can lead to the overlapping of these characteristics (Fig 14).

The situation when such characteristics have any common points is unacceptable. This results in unneeded cuts-off during the normal operation of distribution network. Unneeded cuts-off of highly loaded lines lead to increases of loads of adjacent lines and cascading failures potentially culminating in blackouts.

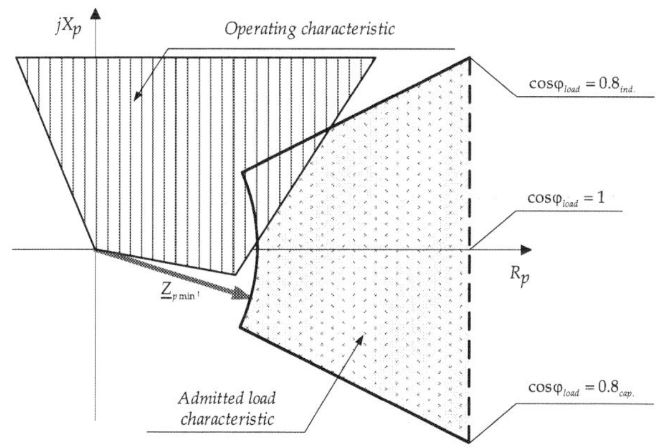

Fig. 14. Overlapping of operating and admitted load characteristics

The impedance area covering the admitted loads of a power line is dependent on the level and the character of load. This means that the variable parameters are both the amplitude and the phase part of the impedance vector. In normal operating conditions the amplitude of load impedance changes from Z_{pmin} practically to the infinity (unloaded line). The phase of load usually changes from $\cos\varphi$ = 0.8_{ind} to $\cos\varphi$ = 0.8_{cap}. The expected Z_{pmin} can be determined by the following equation (Ungrad et al., 1995), (Schau et al., 2008):

$$Z_{p\min} = \frac{U_{p\min}^2}{S_{p\max}} = \frac{U_{p\min}}{\sqrt{3}\cdot I_{p\max}}, \tag{13}$$

where:

U_{pmin} - minimal admitted operating voltage in kV (usually U_{pmin} = 0,9 U_N),
S_{pmax} - maximum apparent power in MVA,
I_{pmax} - maximum admitted load.

A necessary condition of connecting DPGS to the HV network is researching whether the increase of load (especially in faulty conditions e.g. one of the lines is falling out) is not leading to an overlap. Because of the security reasons and the falsifying factors influencing the impedance evaluation, it is assumed that the protection will not unnecessarily pick-up if the impedance reach of operating zones will be shorter than 80% of the minimal expected load. This requirement will be practically impossible to meet especially when the MHO starting characteristics are used (Fig 15a). There are more possibilities when the protection realizes a distance protection function with polygonal characteristics (Fig. 15b).

Using digital distance protections with polygonal characteristics is also very effective for HV lines equipped with high temperature low sag conductors or thermal line rating. In this case the load can increase 2.5 times. Figure 16 shows the adaptation of an impedance area to the maximum expected power line load. Of course this implies serious problems with the recognition of faults with high resistances.

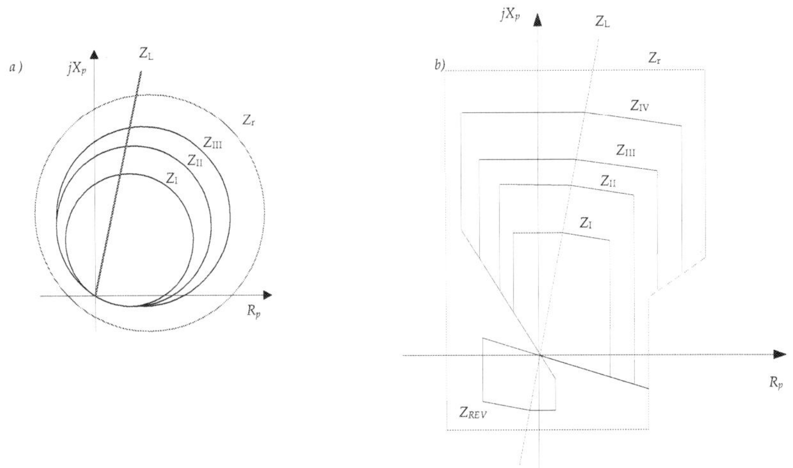

Fig. 15. Starting and operating characteristics a) MHO, b) polygonal

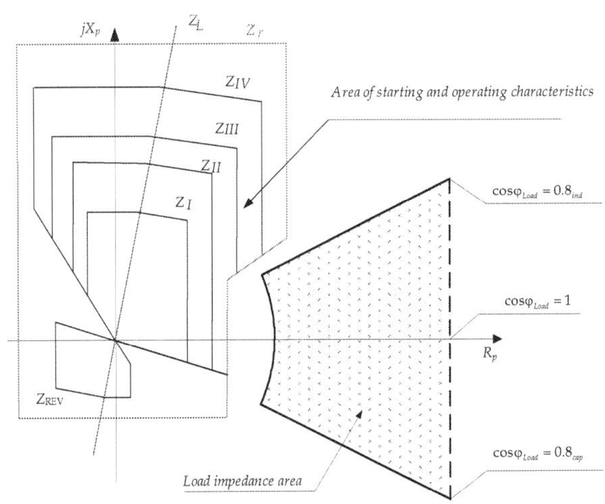

Fig. 16. Adaptation of operating characteristics to the load impedance area

5.2 Simulations

Figure 17 shows the network structure taken for the determination of the influence of selected factors on the impedance evaluation error. This is a part of the 110 kV network of the following parameters:

- short-circuit powers of equivalent systems: $S''_{kA} = 1000$ MVA, $S''_{kB} = 500$ MVA;
- wind farm consists of 30 wind turbines using double fed induction generators of the individual power P_{jN}=2 MW with a fault ride-through function. Power of a wind farm is changing from 10% to 100% of the nominal power of the wind farm. WF is connected in the three-terminal line scheme,
- overhead power line AB:
- length: 30 km; resistance per km: r_l=0.12 Ω/km, reactance per km x_j=0.4 Ω/km
- overhead power output line from WF:
- length: 2 km; resistance per km: r_l=0.12 Ω/km, reactance per km x_j=0.4 Ω/km
- metallic three-phase fault on line AB between the M connection point and 100% of the line $L_{A\text{-}B}$ length.

Initial and steady fault currents from the wind farm and system A have been evaluated for these parameters. It has been assumed that phases of these currents are equal. The initial fault current of individual wind turbines will be limited to 330% of the nominal current of the generator and wind turbines will generate steady fault current on the level of 110% of the nominal current of the generator. The following examples will now be considered.

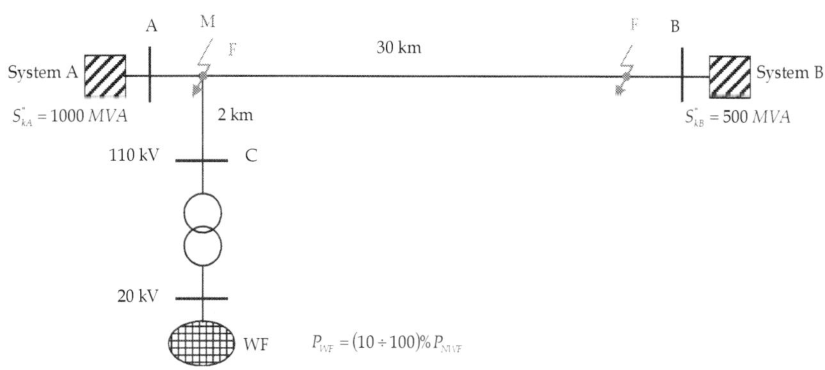

Fig. 17. Network scheme for simulations

Example 1

The network is operating in quasi-steady conditions. The farm is generating power of 60 MW and is connected at 10 % of the $L_{A\text{-}B}$ line length. The location of a fault changeable from 20 % to 100 % of the $L_{A\text{-}B}$ length with steps of 10 %. Table 1 presents selected results of simulations for faults of times not exceeding 50 ms. Results take into consideration the limitation of fault currents on the level of 330% of the nominal current of the generator. By analogy, Table 2 shows the results when the limitation is 110 % after a reaction of the control units.

Fault location		I_A	I_C	I_C/I_A	ΔR	ΔX	$\delta_{R\%}$	$\delta_{X\%}$	R_{LAF}	X_{LAF}
l	$x_{\%}\underline{Z}_{LAB}$									
[km]	[%]	[kA]	[kA]	[-]	[Ω]	[Ω]	[%]	[%]	[Ω]	[Ω]
6	20	3.93	0.801	0.204	0.073	0.245	10.191	10.191	0.720	2.400
9	30	3.591	0.732	0.204	0.147	0.489	13.590	13.590	1.080	3.600
12	40	3.305	0.674	0.204	0.220	0.734	15.295	15.295	1.440	4.800
15	50	3.061	0.624	0.204	0.294	0.979	16.308	16.308	1.800	6.000
18	60	2.851	0.581	0.204	0.367	1.223	16.982	16.982	2.160	7.200
21	70	2.667	0.545	0.204	0.441	1.471	17.516	17.516	2.520	8.400
24	80	2.505	0.511	0.204	0.514	1.714	17.849	17.849	2.880	9.600
27	90	2.362	0.481	0.204	0.586	1.955	18.101	18.101	3.240	10.800
30	100	2.234	0.455	0.204	0.660	2.200	18.330	18.330	3.600	12.000

Table 1. Initial fault currents and impedance errors for protection located in station A depending on the distance to the location of a fault (Case 1)

where:
l - distance to a fault from station A,
$x_{\%}\underline{Z}_{LAB}$ - distance to a fault in the percentage of the L_{AB} length,
I_A - rms value of the initial fault current flowing from system A to the point of fault,
I_C - rms value of the initial current flowing from WF to the point of a fault,
ΔR - absolute error of the resistance evaluation of the impedance algorithm, $\Delta R = \mathrm{Re}\{(I_C/I_A)\,\underline{Z}_{LMF}\}$,
ΔX - absolute error of the reactance evaluation of the impedance algorithm, $\Delta X = \mathrm{Im}\{(I_C/I_A)\,\underline{Z}_{LMF}\}$,
R_{LAF} - real value of the resistance of the fault loop,
X_{LAF} - real value of the reactance of the fault loop,
$\delta_{R\%}$ - relative error of the evaluation of the resistance $\delta_{R\%} = \Delta R/R_{LAF}$,
$\delta_{X\%}$ - relative error of the evaluation of the resistance, $\delta_{X\%} = \Delta X/X_{LAF}$.

Fault location		$I_{A(u)}$	$I_{C(u)}$	$I_{C(u)}/I_{A(u)}$	ΔR	ΔX	$\delta_{R\%}$	$\delta_{X\%}$
l	$x_{\%}\underline{Z}_{LAB}$							
[km]	[%]	[kA]	[kA]	[-]	[Ω]	[Ω]	[%]	[%]
6	20	3.986	0.328	0.082	0.030	0.099	4.114	4.114
9	30	3.685	0.328	0.089	0.064	0.214	5.934	5.934
12	40	3.425	0.328	0.096	0.103	0.345	7.182	7.182
15	50	3.199	0.328	0.103	0.148	0.492	8.203	8.203
18	60	3	0.328	0.109	0.197	0.656	9.111	9.111
21	70	2.824	0.328	0.116	0.251	0.836	9.955	9.955
24	80	2.666	0.328	0.123	0.310	1.033	10.765	10.765
27	90	2.525	0.328	0.130	0.374	1.247	11.547	11.547
30	100	2.398	0.328	0.137	0.443	1.477	12.310	12.310

Table 2. Steady fault currents and impedance errors for protection located in station A depending on the distance to the location of a fault (Case 2)

where:

$I_{A(u)}$ - rms value of steady fault current flowing from system A to the point of a fault,

$I_{C(u)}$ - rms value of steady fault current flowing from WF to the point of a fault,

The above-mentioned tests confirm that the presence of sources of constant generated power (WF) brings about the miscalculation of impedance components. The error is rising with the distancing from busbars in substation A to the point of a fault, but does not exceed 20 %. It can be observed at the beginning of a fault that the error level is higher than in the case of action of the wind farm control units. It is directly connected with the quotient of currents from system A and WF. In the first case it is constant and equals 0.204. In the second one it is lower but variable and it is rising with the distance from busbars of substation A to the point of a fault.

From the point of view of distance protection located in station C powered by WF, the error level of evaluated impedance parameters is much higher and exceeds 450 %. It is due to the high I_A/I_C ratio which is 4.9. Figure 18 shows a comparison of a relative error of estimated reactance component of the impedance fault loop for protection located in substation A (system A) and station C (WF).

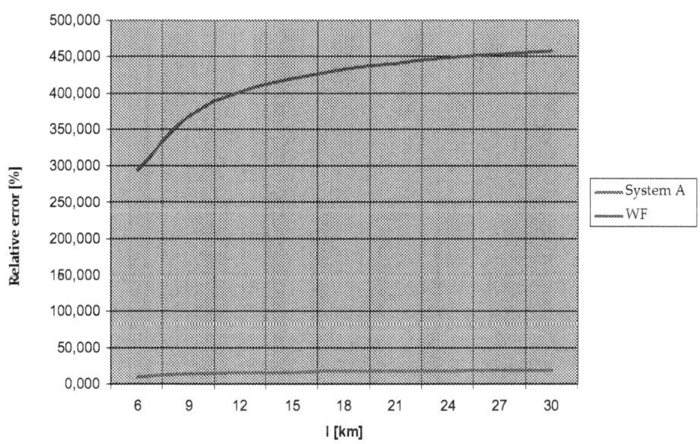

Fig. 18. Relative error (%) of reactance estimation in distance protection in substation A and C in relation to the distance to a fault

Attempting to compare estimates of impedance components for distance protections in substations A, B and C in relation to the distance to a fault, the following analysis has been undertaken for the network structure as in Fig. 19. Again a three-terminal line of WF connection has been chosen as the most problematic one for power system protections. For this variant WF consists of 25 wind turbines equipped as before with DFIG generators each of 2 MW power. The selection of such a type of generator is dictated by its high fault currents when compared with generators with power converters in the power output path and the popularity of the first ones.

Figure 20 shows the influence of the location of a fault on the divergence of impedance components evaluation in substations A, B and C in comparison to the real expected values. The presented values are for the initial time of a three-phase fault on line A-B with the assumption that all wind turbines are operating simultaneously, generating the nominal power.

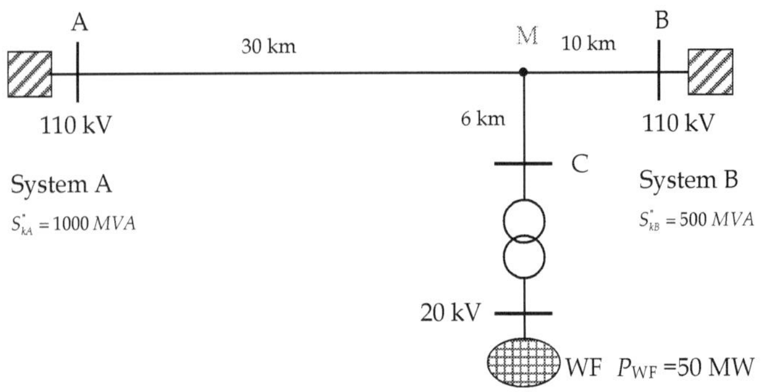

Fig. 19. Network scheme for the second stage of simulations

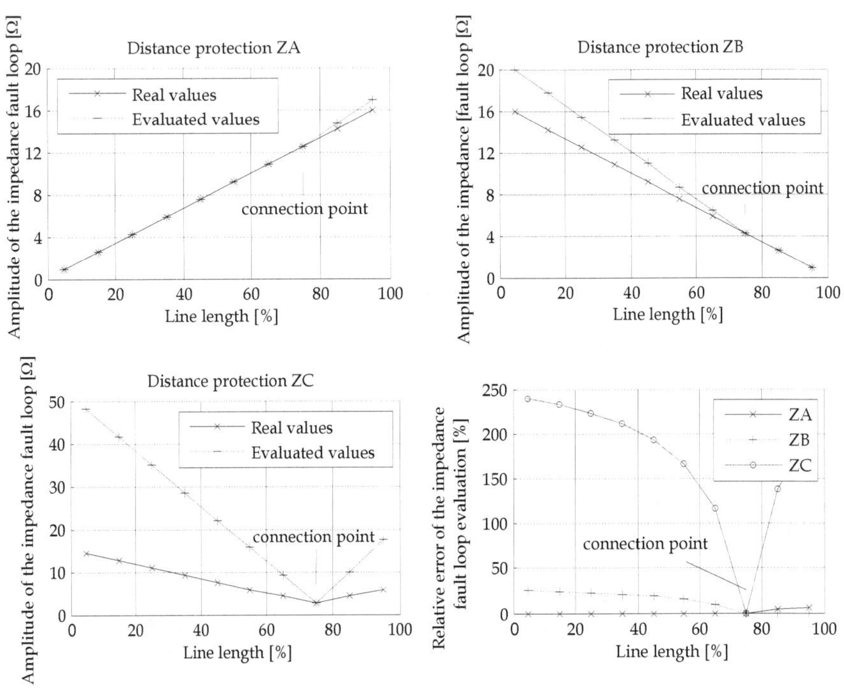

Fig. 20. Divergences between the evaluated and expected values of the amplitude of impedance for protections in substations A, B and C

Analyzing courses in Fig. 20, it can be observed that the highest inaccuracy in the amplitude of impedance evaluation concerns protections in substation C. The divergences between evaluated and expected values are rising along with the distance from the measuring point to the location of fault. It is characteristic that in substations A and B these divergences are at least one class lower than for substation C. This is the consequence of a significant

disproportion of the short-circuit powers of systems A and B in relation to the nominal power of WF.
On the other hand, for the fault in the C-M segment of line the evaluation error of an impedance fault loop is rising for distance protections in substations A and B. For distance protection in substation B a relative error is 53 % at fault point located 4 km from the busbars of substation C. For distance of 2 km from station C the error exceeds 86 % of the real impedance to the location of a fault (Lubośny, 2003).

Example 2

The network as in Figure 17 is operating with variable generating power of WF from 100 % to 10 % of the nominal power. The connection point is at 10 % of the line $L_{A\text{-}B}$ length. A simulated fault is located at 90 % of the $L_{A\text{-}B}$ length.
Table 3 shows the initial fault currents and error levels of estimated impedance components of distance protections in stations A and C. Changes of WF generating power P_{WF} influence the miscalculations both for protections in station A and C. However, what is essential is the level of error. For protection in station A the maximum error level is 20 % and can be corrected by the modification of reactance setting by 2 Ω (when the reactance of the line L_{AB} is 12 Ω). This error is dropping with the lowering of the WF generated power (Table 3).

WF power		I''_{kA}	I''_{kC}	$\delta_{R(A)\%}$	$\delta_{X(A)\%}$	$\delta_{R(C)\%}$	$\delta_{X(C)\%}$
P_{WF}	$_{\%}P_{WFN}$						
[MW]	[%]	[kA]	[%]	[%]	[%]	[%]	[%]
60	100	2.362	0.481	18.101	18.101	453.286	453.286
54	90	2.374	0.453	16.962	16.962	483.749	483.749
48	80	2.386	0.422	15.721	15.721	521.910	521.910
42	70	2.401	0.388	14.364	14.364	571.213	571.213
36	60	2.416	0.35	12.877	12.877	637.187	637.187
30	50	2.433	0.308	11.253	11.253	729.171	729.171
24	40	2.454	0.261	9.454	9.454	867.905	867.905
18	30	2.474	0.208	7.473	7.473	1097.929	1097.929
12	20	2.499	0.148	5.264	5.264	1558.628	1558.628
6	10	2.527	0.079	2.779	2.779	2952.678	2952.678

Table 3. Initial fault currents and relative error levels of impedance estimation for protections in substations A and C in relation to the WF generated power

For protection in substation C the error level is rising with the lowering of WF generated power. Moreover the level of this error is several times higher than for protection in station A. The impedance correction should be ΔR−92.121 Ω and ΔX−307.078 Ω. For the impedance of L_{CB} segment $\underline{Z}_{LCB}$=(3.48+j11.6) Ω such correction is practically impossible. With this correction the impedance reach of operating characteristics of distance protections in substation C will be deeply in systems A and B. Figure 21 shows the course of error level of estimated resistance and reactance in protections located in the substations A and C in relation to the WF generated power.
When the duration of a fault is so long that the control units of WF are coming into action, the error level of impedance components evaluation for protections in the station C is still rising. This is the consequence of the reduction of WF participation in the total fault current.

Figure 22 shows the change of the quotient of steady fault currents flowing from substations A and C in relation to WF generated power P_{WF}.

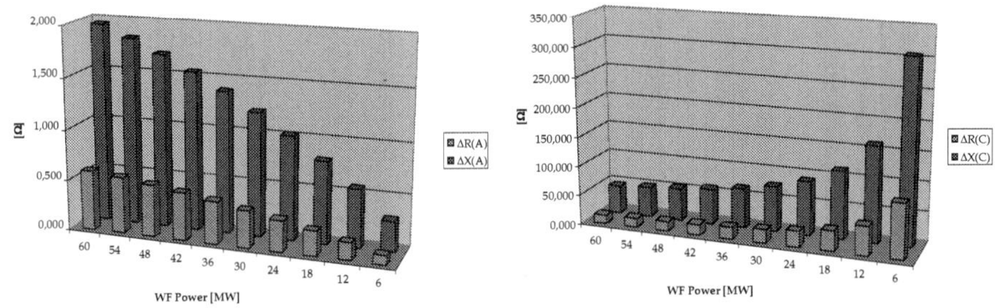

Fig. 21. Impedance components estimation errors in relation to WF generated power for protections a) in substation A, b) in substation C

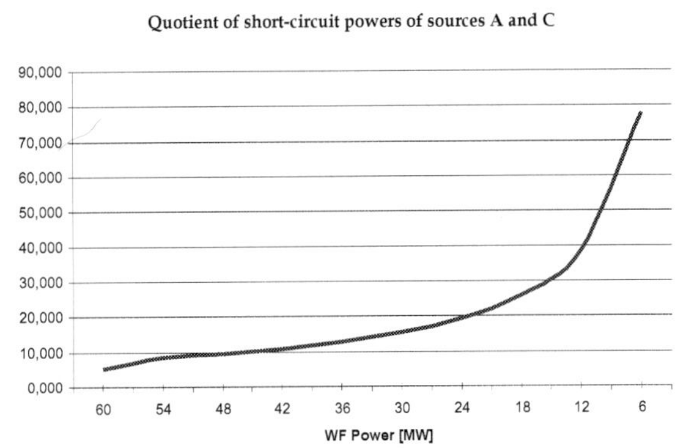

Fig. 22. Change of the quotient of steady fault currents flowing from sources B and C in relation of WF generated power

Example 3

Once again the network is operating as in Figure 17. There are quasi-steady conditions, WF is generating the nominal power of 60 MW, the fault point is at 90 % of the LA-B length. The changing parameter is the location of WF connection point. It is changing from 3 to 24 km from substation A.

Also for these conditions a higher influence of WF connection point location on the proper functioning of power protections can be observed in substation C than in substations A and B. The further the connection point is away from substation A, the lower are the error levels of estimated impedance components in substations A and C. It is the consequence of the rise of WF participation in the initial fault current (Table 4). The error levels for protections in substation A are almost together, whereas in substation C they are many times lower than in the case of a change in the WF generated power. If the fault time is so long that the control units of WF will come into action, limiting the WF fault current, the error level for protections in substation C will rise more. This is due to the quotient $I_{A(u)}/I_{C(u)}$ which is leading to the rise of estimation error $\underline{\Delta Z}_{(C)} = \underline{Z}_{MF}\frac{\underline{I}_{A(u)}}{\underline{I}_{C(u)}}$.

Figure 23 shows the course of error of reactance estimation for the initial and steady fault current for impedances evaluated by the algorithms implemented in protection in substation C.

WF connection point location	I_A	I_C	I_C/I_A	I_A/I_C	$\Delta R_{(A)}$	$\Delta X_{(A)}$	$\Delta R_{(C)}$	$\Delta X_{(C)}$
[km]	[kA]	[kA]	[-]	[-]	[Ω]	[Ω]	[Ω]	[Ω]
3	2.362	0.481	0.204	4.911	0.586	1.955	14.143	47.142
6	2.371	0.525	0.221	4.516	0.558	1.860	11.381	37.936
9	2.385	0.57	0.239	4.184	0.516	1.721	9.038	30.126
12	2.402	0.617	0.257	3.893	0.462	1.541	7.007	23.358
15	2.424	0.6652	0.274	3.644	0.395	1.317	5.247	17.491
18	2.45	0.716	0.292	3.422	0.316	1.052	3.696	12.318
21	2.48	0.769	0.310	3.225	0.223	0.744	2.322	7.740
24	2.518	0.825	0.328	3.052	0.118	0.393	1.099	3.663

Table 4. Values and quotients of the initial fault currents flowing from sources A and C, and the error levels of impedance components estimation in relation to the WF connection point location

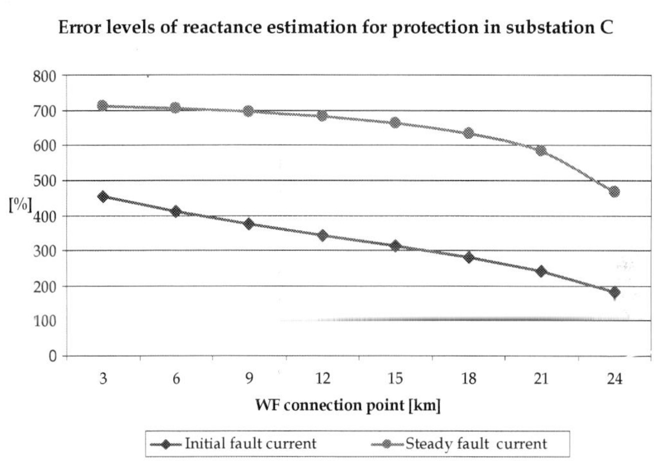

Fig. 23. Error level of the reactance estimation for distance protection in substation C in relation of WF connection point

Taking the network structure shown in Fig. 24, according to distance protection principles, the reach of the first zone should be set at 90 % of the protected line length. But in this case, if the first zone is not to reach the busbars of the surrounding substations, the maximum reactance settings should not exceed:

For distance protection in substation A: $X_{1A} < (1.2 + 0.8) = 2\,\Omega$

For distance protection in substation B: $X_{1B} < (10.8 + 0.8) = 11.6\,\Omega$

For distance protection in substation C: $X_{1C} < (1.2 + 0.8) = 2\,\Omega$

With these settings most of the faults on segment L_{MB} will not be switched off with the self-time of the first zone of protection in substation A. This leads to the following switching-off sequence. The protection in substation B will switch off the fault immediately. The network will operate in configuration with two sources A and C. If the fault has to be switched off with the time Δt, the reaches of second zones of protections in substations A and C have to include the fault location. So their reach must extend deeply into the system A and the WF structure. Such a solution will produce serious problems with the selectivity of functioning of power protection automation.
Taking advantage of the in-feed factor k_{if} also leads to a significant extension of these zones, especially for protection in substation C. Due to the highly changeable value of this factor in relation to the WF generated power and the location of connection, what will be efficient is only adaptive modified settings, according to the operating conditions identified in real time.

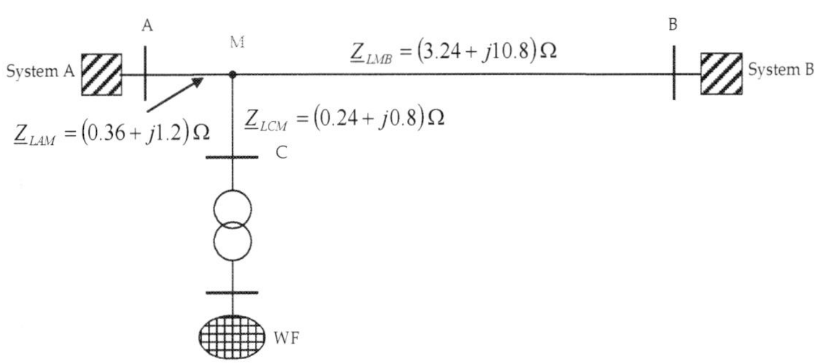

Fig. 24. Simplified impedance scheme of the network structure from the Figure 17

6. Conclusions

The presented selected factors influencing the estimation of impedance components in digital protections, necessitate working out new protection structures. These must have strong adaptive abilities and the possibility of identification, in real time, of an actual operating state (both configuration of interconnections and parameters of work) of the network structure. The presented simulations confirm that the classic parameterization of distance protections, even the one taking into account the in-feed factor k_{if} does not yield effective and selective fault eliminations.
Nowadays distance protections have individual settings for the resistance and reactance reaches. Thus the approach of the resistance reach and admitted load area have to be taken

into consideration. Resistance reach should include faults with an arc and of high resistances. This is at odds with the common trend of using high temperature low sag conductors and the thermal line rating, which of course extends the impedance area of admitted loads. As it has been shown, also the time of fault elimination is the problem for distance protections in substations in the WF surrounding, when this time is so long that the WF fault current is close to their nominal current value.

Simulation results prove that the three-terminal line type of DPGS connection, especially wind farms, to the distribution network contributes to the significant shortening of the reaches of distance protections. The consequences are:

- extension of fault elimination time (switching off will be done with the time of the second zone instead of the self-time first zone),
- incorrectness of autoreclosure automation functioning (e.g. when in the case of shortening of reaches the extended zones will not include the full length of line),
- no reaction of protections in situations when there is a fault in the protected area (missing action of protection) or delayed cascaded actions of protections.

A number of factors influencing the settings of distance protections, with the presence of wind farms, causes that using these protections is insufficient even with pilot lines. So new solutions should be worked out. One of them is the adaptive area automation system. It should use the synchrophasors technique which can evaluate the state estimator of the local network, and, in consequence, activates the adapted settings of impedance algorithms to the changing conditions. Due to the self-time of the first zones (immediate operation) there is a need for operation also in the area of individual substations. Thus, it is necessary to work out action schemes in the case of losing communication within the dispersed automation structure.

7. References

Datasheet: Vestas, Advance Grid Option 2, V52-850 kW, V66-1,75 MW, V80-2,0 MW, V90-1,8/2,0 MW, V90-3,0 MW.

Halinka, A.; Sowa, P. & Szewczyk M. (2006): Requirements and structures of transmission and data exchange units in the measurement-protection systems of the complex power system objects. *Przegląd Elektrotechniczny (Electrical Review)*, No. 9/2006, pp. 104 - 107, ISSN 0033-2097 (in Polish)

Halinka, A. & Szewczyk, M. (2009): Distance protections in the power system lines with connected wind farms, *Przegląd Elektrotechniczny (Electrical Reviev)*, R 85, No. 11/2009, pp. 14 - 20, ISSN 0033-2097 (in Polish)

Lubośny, Z. (2003): *Wind Turbine Operation in Electric Power Systems. Advanced Modeling*, Springer-Verlag, ISBN: 978-3-540-40340-1, Berlin Heidelberg New York

Pradhan, A. K. & Geza, J. (2007): Adaptive distance relay setting for lines connecting wind farms. *IEEE Transactions on Energy Conversion*, Vol 22, No.1, March 2007, pp. 206-213

Shau, H.; Halinka, A. & Winkler, W. (2008): *Elektrische Schutzeinrichtungen in Industrienetzen und -anlagen. Grundlagen und Anwendungen*, Hüting & Pflaum Verlag GmbH & Co. Fachliteratur KG, ISBN 978-3-8101-0255-3, München/Heidelberg (in German)

Ungrad, H.; Winkler, W. & Wiszniewski A. (1995): *Protection techniques in Electrical Energy Systems*, Marcel Dekker, Inc., ISBN 0-8247-9660-8, New York

Ziegler, G. (1999): *Numerical Distance Protection. Principles and Applications, Publicis MCD,* ISBN 3-89578-142-8

8

Impact of Intermittent Wind Generation on Power System Small Signal Stability

Libao Shi[1], Zheng Xu[1], Chen Wang[1], Liangzhong Yao[2] and Yixin Ni[1]
[1]*Graduate School at Shenzhen, Tsinghua University Shenzhen 518055,*
[2]*Alstom Grid Research & Technology Centre, Stafford, ST17 4LX,*
[1]*China*
[2]*United Kingdom*

1. Introduction

In recent years, the increasing concerns to environmental issues demand the search for more sustainable electrical sources. Wind energy can be said to be one of the most prominent renewable energy sources in years to come (Ackermann, 2005). And wind power is increasingly considered as not only a means to reduce the CO_2 emissions generated by traditional fossil fuel fired utilities but also a promising economic alternative in areas with appropriate wind speeds. Albeit wind energy currently supplies only a fraction of the total power demand relative to the fossil fuel fired based conventional energy source in most parts of the world, statistical data show that in Northern Germany, Denmark or on the Swedish Island of Gotland, wind energy supplies a significant amount of the total energy demand. Specially it should be pointed out that in the future, many countries around the world are likely to experience similar penetration levels. Naturally, in the technical point of view, power system engineers have to confront a series of challenges when wind power is integrated with the existing power system. One of important issues engineers have to face is the impact of wind power penetration on an existing interconnected large-scale power system dynamic behaviour, especially on the power system small signal stability. It is known that the dynamic behavior of a power system is determined mainly by the generators. So far, nearly all studies on the dynamic behavior of the grid-connected generator under various circumstances have been dominated by the conventional synchronous generators world, and much of what is to be known is known. Instead, the introduction of wind turbines equipped with different types of generators, such as doubly-fed induction generator (DFIG), will affect the dynamic behaviour of the power system in a way that might be different from the dominated synchronous generators due to the intermittent and fluctuant characteristics of wind power in nature. Therefore, it is necessary and imperative to study the impact of intermittent wind generation on power system small signal stability.

It should be noticed that most published literature are based on deterministic analysis which assumes that a specific operating situation is exactly known without considering and responding to the uncertainties of power system behavior. This significant drawback of deterministic stability analysis motivates the research of probabilistic stability analysis in which the uncertainty and randomness of power system can be fully understood. The probabilistic stability analysis method can be divided into two types: the analytical method, such as point estimate method (Wang et al., 2001); and the simulation method, such as Monte Carlo Simulation (Rueda et al., 2009). And most published literature related to probabilistic stability analysis are based on the uncertainty of traditional generators with simplified probability distributions. With increasing penetration levels of wind generation, and considering that the uncertainty is the most significant characteristic of wind generation, a more comprehensive probabilistic stability research that considering the uncertainties and intermittence of wind power should be conducted to assess the influence of wind generation on the power system stability from the viewpoint of probability.

Generally speaking, the considered wind generation intermittence is caused by the intermittent nature of wind source, i.e. the wind speed. Correspondingly, the introduction of the probability distribution of the wind speed is the key of solution. In our work, the well-known Weibull probability density function for describing wind speed uncertainty is employed. In this chapter, according to the Weibull distribution of wind speed, the Monte Carlo simulation technique based probabilistic small signal stability analysis is applied to solve the probability distributions of wind farm power output and the eigenvalues of the state matrix.

2. Wind turbine model

In modelling turbine rotor, there are a lot of different ways to represent the wind turbine. Functions approximation is a way of obtaining a relatively accurate representation of a wind turbine. It uses only a few parameters as input data to the turbine model. The different mathematical models may be more or less complex, and they may involve very different mathematical approaches, but they all generate curves with the same fundamental shapes as those of a physical wind turbine.

In general, the function approximations representing the relation between wind speed and mechanical power extracted from the wind given in Equation (1) (Ackermann, 2005) are widely used in modeling wind turbine.

$$P_m = \begin{cases} 0 & V_w \le V_{cut-in} \\ 0.5 \cdot \rho \cdot A_{wt} \cdot C_p(\beta, \lambda) \cdot V_w^{\ 3} & V_{cut-in} < V_w \le V_{rated} \\ p_r & V_{rated} < V_w < V_{cut-off} \\ 0 & V_w \ge V_{cut-off} \end{cases} \tag{1}$$

where P_m is the power extracted from the wind; ρ is the air density; C_p is the performance coefficient; λ is the tip-speed ratio (v_t/v_w), the ratio between blade tip speed, v_t (m/s), and wind speed at hub height upstream of the rotor, v_w (m/s); $A_{wt}=\pi R^2$ is the area covered by the wind turbine rotor, R is the radius of the rotor; V_w denotes the wind speed; and β is the blade pitch angle; V_{cut-in} and $V_{cut-offt}$ are the cut-in and cut-off wind speed of wind turbine; V_{rated} is the wind speed at which the mechanical power output will be the rated power. When V_w is higher than V_{rated} and lower than $V_{cut-off}$, with a pitch angle control system, the mechanical power output of wind turbine will keep constant as the rated power.

It is known that the performance coefficient C_p is not a constant. Usually the majority of wind turbine manufactures supply the owner with a C_p curve. The curve expresses C_p as a function of the turbine's tip-speed ratio λ. However, for the purpose of power system stability analysis of large power systems, numerous researches have shown that C_p can be assumed constant. Fig. 1 (Akhmatov, 2002) gives the curves of performance coefficient C_p with changing of rotational speed of wind turbine at different wind speed conditions ($\beta$is fixed). According to Fig. 1, by adjusting the rotational speed of the rotor to its optimized value $\omega_{m\text{-opt}}$, the optimal performance coefficient $C_{p\max}$ can be reached.

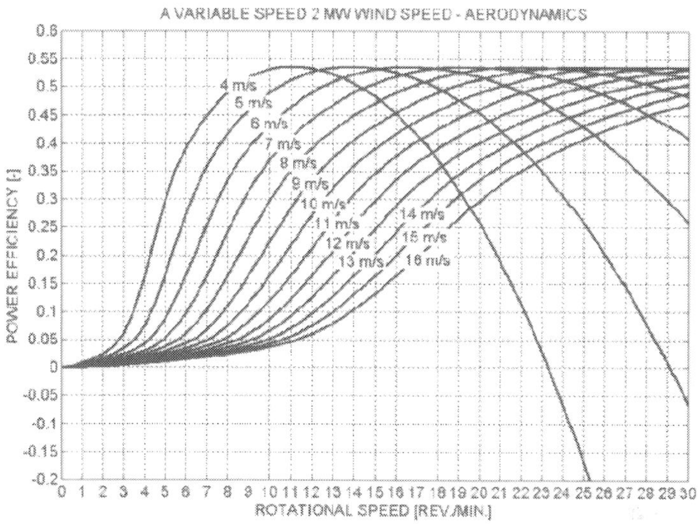

Fig. 1. Curves of C_p with changing of ω_m at different wind speed

In this chapter, we assume that for any wind speed at the range of $V_{cut\text{-}in} < V_w \leq V_{rated}$, the rotational speed of rotor can be controlled to its optimized value, therefore the $C_{p\max}$ can be kept constant.

3. Mathematical model of DFIG

The configuration of a DFIG, with corresponding static converters and controllers is given in Fig.1. Two converts are connected between the rotor and grid, following a back to back scheme with a dc intermediate link. Fig.2 gives the reference frames, where a, b and c indicate stator phase a, b and c winding axes; A, B and C indicate rotor phase A, B and C winding axes, respectively; x-y is the synchronous rotation coordinate system in the grid side; θ is the angle between q axis and x axis.

Applying Park's transformation, the voltage equations of a DFIG in the d-q coordinate system rotating at the synchronous speed ω_s, in accordance with generator convention, which means that the stator and rotor currents are positive when flowing towards the network, and real and reactive powers are positive when fed into grid, can be deducted as follows in a per unit system.

$$U_{ds} = -R_s I_{ds} - \psi_{qs} + \frac{1}{\omega_s}\frac{d\psi_{ds}}{dt} \tag{2}$$

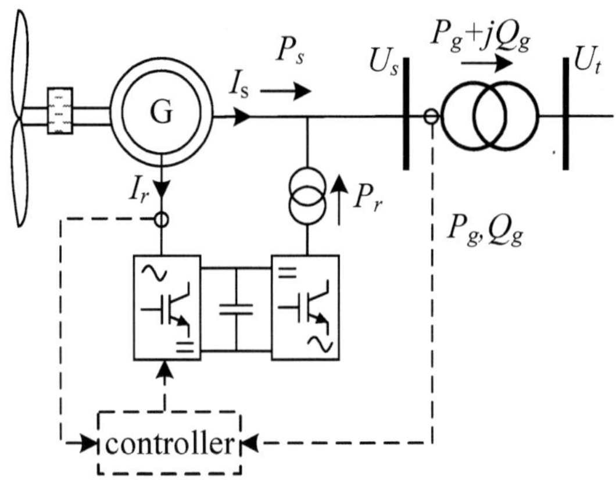

Fig. 2. Schematic diagram of DFIG with converters and controllers

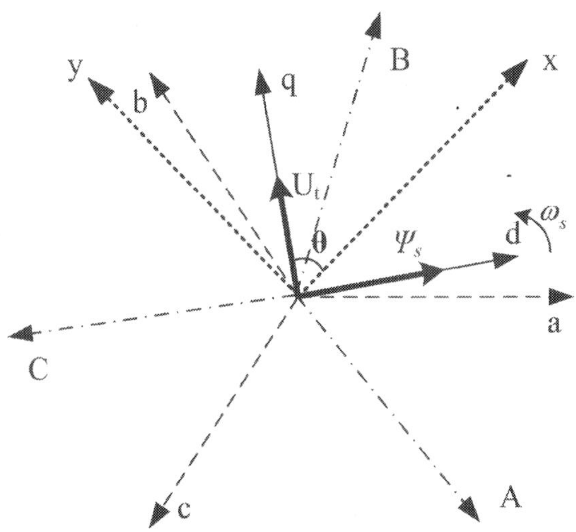

Fig. 3. Reference coordinates for DFIG

$$U_{qs} = -R_s I_{qs} + \psi_{ds} + \frac{1}{\omega_s}\frac{d\psi_{qs}}{dt} \tag{3}$$

$$U_{dr} = -R_r I_{dr} - s\psi_{qr} + \frac{1}{\omega_s}\frac{d\psi_{dr}}{dt} \tag{4}$$

$$U_{qr} = -R_r I_{qr} + s\psi_{dr} + \frac{1}{\omega_s}\frac{d\psi_{qr}}{dt} \tag{5}$$

$$\psi_{ds} = -X_s I_{ds} - X_m I_{dr} \tag{6}$$

$$\psi_{qs} = -X_s I_{qs} - X_m I_{qr} \tag{7}$$

$$\psi_{dr} = -X_r I_{dr} - X_m I_{ds} \tag{8}$$

$$\psi_{qr} = -X_r I_{qr} - X_m I_{qs} \tag{9}$$

$$P_g = P_s + P_r = (U_{ds} I_{ds} + U_{qs} I_{qs}) + (U_{dr} I_{dr} + U_{qr} I_{qr}) \tag{10}$$

$$Q_g = Q_s + Q_r = (U_{qs} I_{ds} - U_{ds} I_{qs}) + (U_{qr} I_{dr} - U_{dr} I_{qr}) \tag{11}$$

$$2H\frac{ds}{dt} = T_e - T_m \tag{12}$$

Where U, I, Ψ denote the voltage, current and flux linkage; P and Q denote the real and reactive power outputs of wind generator, respectively; T_m and T_e denote the mechanical and electromagnetic torques of wind generator, respectively; R and X denote resistance and reactance, respectively; the subscripts r and s denote the stator and rotor windings, respectively; the subscript g means generator; H is the inertia constant, and t stands for time; s is the slip of speed.

The reactances X_s and X_r can be calculated in following equations.

$$X_s = X_{s\sigma} + X_m \tag{13}$$

$$X_r = X_{r\sigma} + X_m \tag{14}$$

Where $X_{s\sigma}$ and $X_{r\sigma}$ are the leakage reactances of stator and rotor windings, respectively; X_m is the mutual reactance between stator and rotor.

The aforementioned equations describe the electrical dynamic performance of a wind turbine, namely, the asynchronous machine. However, these equations are not suitable for small signal analysis directly. It is necessary and imperative to deduce the simplified and practical model. The following assumptions are presented to model the DFIG.

a. Magnetic saturation phenomenon is not considered during modelling;
b. For the wind turbine equipped with DFIG, all rotating masses are represented by one element, which means that a so-called 'lumped-mass' or 'one-mass' representation is used;
c. The stator transients and stator resistance are negligible, i.e. $\frac{d\psi_{ds}}{dt} = 0$, $\frac{d\psi_{qs}}{dt} = 0$, and R_s=0 in Eqs (2) and (3).

Furthermore, the stator flux-oriented control strategy (Tapia et al., 2006) is adopted in this work, which makes the stator flux ψ_s line in accordance with d-axis, as depicted in Fig.3., i.e.

$$\psi_{ds} = \psi_s \tag{15}$$

$$\psi_{qs} = 0 \tag{16}$$

Then the stator voltage equations can be rewritten as

$$U_{ds} = 0 \tag{17}$$

$$U_{qs} = \psi_s = U_t \tag{18}$$

Where U_t is the terminal voltage;
From Fig. 3, the vector of stator voltage U_s=U_t is always align with q axis with the stator flux-oriented control strategy. And according to the stator flux linkage equations (6) and (7), the stator currents I_{ds} and I_{qs} can be represented as the function of rotor current and terminal voltage U_t, i.e.

$$I_{ds} = -\frac{1}{X_s}(U_t + X_m I_{dr}) \tag{19}$$

$$I_{qs} = -\frac{1}{X_s} X_m I_{qr} \tag{20}$$

Substituting equations (8) and (9) in equations (4) and (5), we find

$$U_{dr} = -R_r I_{dr} - \frac{X'_r}{\omega_s}\frac{dI_{dr}}{dt} + sX'_r I_{qr} \tag{21}$$

$$U_{qr} = -R_r I_{qr} - \frac{X'_r}{\omega_s}\frac{dI_{qr}}{dt} - sX'_r I_{dr} + s\frac{X_m}{X_s}\psi_s \tag{22}$$

Where $X_r'=X_r-X_m^2/X_s$.
Consider that the grid-side converter of DFIG always operates at unity power factor, i.e. Q_r= 0, the reactive power Q_g is equal to the stator reactive power Q_s, i.e. Q_g=Q_s. In the steady state analysis, in accordance with the expressions of stator power and the rotor power, it can be proved that P_r=-sP_s, and P_g=P_s/(1-s). Accordingly, the real and reactive powers equations and the torque equation can be rewritten as

$$P_g = P_s / (1-s) = -\frac{U_t}{X_s(1-s)} X_m I_{qr} \tag{23}$$

$$Q_g = Q_s = -\frac{U_t}{X_s}(U_t + X_m I_{dr}) \tag{24}$$

$$2H\frac{ds}{dt} = T_e - T_m = (\psi_{ds} I_{qs} + \psi_{qs} I_{ds}) - T_m = -\frac{X_m}{X_s} U_t I_{qr} - T_m \tag{25}$$

Finally, the equations (17-18), (21-22), (23-25) constitute the 3rd order simplified practical DFIG model.

4. Mathematical model of DFIG Converters

As shown in Fig.2, the model of DFIG frequency converter system consists of rotor-side converter, grid-side converter, the dc link and the corresponding converter control. In this chapter, it is assumed that the grid-side converter is ideal and the dc link voltage between the converters is constant during analysis. This decouples the grid-side converter from the rotor-side converter. The rotor-side converter is assumed to be a voltage-controlled current source, and the stator flux-oriented control strategy is employed to implement the decoupled control of the real and reactive power outputs of DFIG. The overall converter control system consists of two cascaded control loops, i.e. the inner control and the outer control. The inner control loop implements the rotor current control, and the outer control loop implements the power control (Tapia et al., 2006).

In order to implement the decoupled control of the real and reactive power outputs of DFIG, two new variables, $\hat{U}_{dr}$, $\hat{U}_{qr}$ are introduced which are defined as:

$$\hat{U}_{dr} = U_{dr} - sX_r' I_{qr} \tag{26}$$

$$\hat{U}_{qr} = U_{qr} + sX_r' I_{dr} - s\frac{X_m}{X_s}\psi_s \tag{27}$$

The newly introduced variables can fully make the dynamics of d and q axes decoupling. Accordingly, the rotor voltage equations can be rewritten as

$$\hat{U}_{dr} = -R_r I_{dr} - \frac{X_r'}{\omega_s}\frac{dI_{dr}}{dt} \tag{28}$$

$$\hat{U}_{qr} = -R_r I_{qr} - \frac{X_r'}{\omega_s}\frac{dI_{qr}}{dt} \tag{29}$$

In this chapter, two special PI controllers are designed to implement the decoupled control of the real and reactive power outputs of DFIG. The block diagrams of rotor-side converter including the inner and outer control loops expressed in d and q axes are given in Fig.4 and Fig.5. In the rotor current control loop, $T_r'=X_r'/R$, T_r' is the time constant of rotor circuit; I_{drref}, I_{qrref} are the rotor current references in d and q axes, respectively; K_2 and T_2 are the control parameters of PI controller. In the power control loop, P_{sref}, Q_{sref} are the real and reactive power references; K_1, T_1 are the control parameters of PI controller. It should be noted that the specific values of K_1, T_1, K_2 and T_2 can be determined through pole placement method (Tapia et al., 2006).

In accordance with Fig. 4, the corresponding stator real power control model can be described as

$$T_1\frac{dI_{qrref}}{dt} = K_1T_1(\frac{dP_s}{dt} - \frac{dP_{sref}}{dt}) - K_1(P_s - P_{sref}) \tag{30}$$

$$T_2\frac{d\hat{U}_{qr}}{dt} + K_2T_2(\frac{dI_{qr}}{dt} - \frac{dI_{qrref}}{dt}) = K_2(I_{qr} - I_{qrref}) \tag{31}$$

$$\frac{T_r'}{\omega_s}\frac{dI_{qr}}{dt} = -I_{qr} - \frac{\hat{U}_{qr}}{R_r} \tag{32}$$

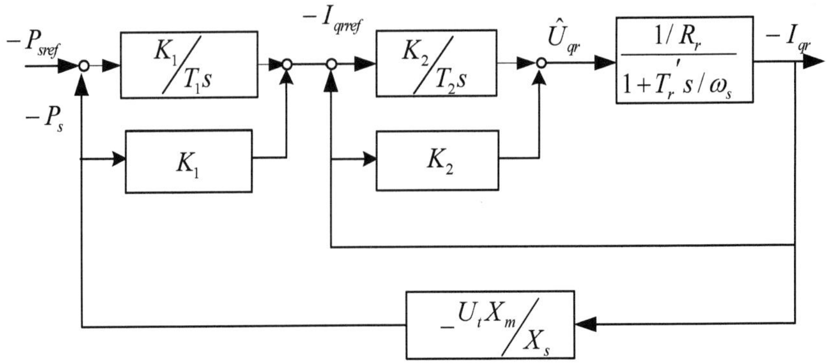

Fig. 4. Block diagram of real power control system in rotor-side converter

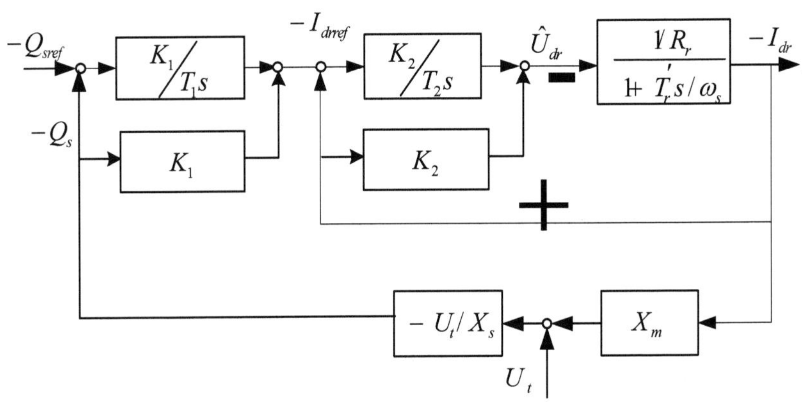

Fig. 5. Block diagram of reactive power control loop in rotor-side converter

$$P_s = -\frac{\psi_s}{X_s} X_m I_{qr} \tag{33}$$

Similarly, the corresponding stator reactive power control model can be described as

$$T_1 \frac{dI_{drref}}{dt} - K_1 T_1 \left(\frac{dQ_s}{dt} - \frac{dQ_{sref}}{dt}\right) = K_1 (Q_s - Q_{sref}) \tag{34}$$

$$T_2 \frac{d\hat{U}_{dr}}{dt} + K_2 T_2 \left(\frac{dI_{dr}}{dt} - \frac{dI_{drref}}{dt}\right) = K_2 (I_{dr} - I_{drref}) \tag{35}$$

$$\frac{T_r'}{\omega_s} \frac{dI_{dr}}{dt} = -I_{dr} - \frac{\hat{U}_{dr}}{R_r} \tag{36}$$

$$Q_s = -\frac{\psi_s}{X_s}(\psi_s + X_m I_{dr}) \tag{37}$$

So far, based on the stator flux-oriented control strategy, and considering the decoupled control of the real and reactive power outputs of DFIG, the whole reduced practical electromechanical transient DFIG model consists of the following 7th order model

$$U_{ds} = 0 \tag{38}$$

$$U_{qs} = U_t \tag{39}$$

$$2H\frac{ds}{dt} = -\frac{X_m}{X_s}U_t I_{qr} - T_m \tag{40}$$

$$\frac{dI_{dr}}{dt} = -\frac{\omega_s}{T_r'}I_{dr} - \frac{\omega_s}{X_r'}\hat{U}_{dr} \tag{41}$$

$$\frac{dI_{drref}}{dt} = K_1 U_{qs}\frac{X_m}{X_s}(\frac{\omega_s}{T_r'} - \frac{1}{T_1})I_{dr} + K_1 U_{qs}\frac{X_m}{X_s}\frac{\omega_s}{X_r'}\hat{U}_{dr} - \frac{K_1}{T_1}(\frac{U_t^2}{X_s} + Q_{sref}) \tag{42}$$

$$\frac{d\hat{U}_{dr}}{dt} = K_2(\frac{1}{T_2} - \frac{\omega_s}{T_r'})I_{dr} - \frac{K_2}{T_2}I_{drref} - K_2\frac{\omega_s}{X_r'}\hat{U}_{dr} \tag{43}$$

$$\frac{dI_{qr}}{dt} = -\frac{\omega_s}{T_r'}I_{qr} - \frac{\omega_s}{X_r'}\hat{U}_{qr} \tag{44}$$

$$\frac{dI_{qrref}}{dt} = K_1 U_{qs}\frac{X_m}{X_s}(\frac{\omega_s}{T_r'} - \frac{1}{T_1})I_{qr} + K_1 U_{qs}\frac{X_m}{X_s}\frac{\omega_s}{X_r'}\hat{U}_{qr} - \frac{K_1}{T_1}P_{sref} \tag{45}$$

$$\frac{d\hat{U}_{qr}}{dt} = K_2(\frac{1}{T_2} - \frac{\omega_s}{T_r'})I_{qr} - \frac{K_2}{T_2}I_{qrref} - K_2\frac{\omega_s}{X_r'}\hat{U}_{qr} \tag{46}$$

5. Model of wind farm of DFIG type

In this chapter, a simple aggregated model of large wind farm in the small signal stability analysis is employed. We assume that currently the operating conditions of all wind generators in a wind farm are same, and the wind farm is considered to be formed with a number of wind generators jointed in parallel. Therefore, the wind farm can be reduced to a single machine equivalent. For a wind farm consisted of N wind generators, the values of stator and rotor voltages are same as the value of single machine. The stator and rotor currents are N times larger than the single machine. The stator and rotor resistances and reactances as well as K_2 are $1/N$ larger than the single machine. The remaining control parameters are same as the single machine.

6. Small signal stability analysis incorporating wind farm of DFIG type

Small signal stability is the ability of the power system to maintain synchronism when subjected to small disturbances (Kundur, 1994). In this context, a disturbance is considered to be small if the equations that describe the resulting response of the system may be linearized for the purpose of analysis. In order to analyze the effects of a disturbance on a linear system, we can observe its eigenvalues. Although power system is nonlinear system, it can be linearized around a stable operating point, which can give a close approximation to the system to be studied.

The behavior of a dynamic autonomous power system can be modelled by a set of n first order nonlinear ordinary differential equations (ODEs) described as follows (Kundur, 1994)

$$\frac{d\mathbf{x}}{dt} = \mathbf{f}(\mathbf{x}, \mathbf{u}) \tag{47}$$

$$0 = \mathbf{g}(\mathbf{x}, \mathbf{u}) \tag{48}$$

where **x** is the state vector; **u** is the vector of inputs to the system; **g** is a vector of nonlinear functions relating state and input variables to output variables.

The equilibrium points of system are those points in which all the derivatives $\dot{x}_1, \dot{x}_2, \ldots, \dot{x}_n$ are simultaneously zero. The system is accordingly at rest since all the variables are constant and unvarying with time. The equilibrium point must therefore satisfy the following equation

$$\frac{d\mathbf{x}_0}{dt} = \mathbf{f}(\mathbf{x}_0, \mathbf{u}_0) = 0 \tag{49}$$

$$0 = \mathbf{g}(\mathbf{x}_0, \mathbf{u}_0) \tag{50}$$

Where ($\mathbf{x}_0$, $\mathbf{u}_0$) are considered as an equilibrium point, which correspond to a basic operating condition of power system.

Corresponding to a small deviation around the equilibrium point, i.e.

$$\mathbf{x} = \mathbf{x}_0 + \Delta\mathbf{x} \tag{51}$$

$$\mathbf{u} = \mathbf{u}_0 + \Delta\mathbf{u} \tag{52}$$

The functions **f**(**x**,**u**) and **g**(**x**,**u**) can be expressed in terms of Taylor's series expansion

$$\frac{d\mathbf{x}_0}{dt} + \frac{d\Delta\mathbf{x}}{dt} = \mathbf{f}(\mathbf{x}_0, \mathbf{u}_0) + \mathbf{A}\Delta\mathbf{x} + \mathbf{B}\Delta\mathbf{u} + \mathbf{O}(\|\Delta\mathbf{x}, \Delta\mathbf{u}\|^2) \tag{53}$$

$$\dot{\mathbf{u}}_0 + \Delta\dot{\mathbf{u}} = \mathbf{g}(\mathbf{x}_0, \mathbf{u}_0) + \mathbf{C}\Delta\mathbf{x} + \mathbf{D}\Delta\mathbf{u} + \mathbf{O}(\|\Delta\mathbf{x}, \Delta\mathbf{u}\|^2) \tag{54}$$

With terms involving second and higher order powers in Eqs(53-54) neglected, we have

$$\Delta\dot{\mathbf{x}} = \mathbf{A}\Delta\mathbf{x} + \mathbf{B}\Delta\mathbf{u} \tag{55}$$

$$0 = \mathbf{C}\Delta\mathbf{x} + \mathbf{D}\Delta\mathbf{u} \tag{56}$$

Where **A**, **B**, **C** and **D** are called as Jacobian matrices represented in the following

$$\mathbf{A} = \left.\frac{\partial \mathbf{f}(\mathbf{x},\mathbf{u})}{\partial \mathbf{x}}\right|_{\mathbf{x}_0,\mathbf{u}_0} \quad (57)$$

$$\mathrm{B} = \left.\frac{\partial \mathbf{f}(\mathbf{x},\mathbf{u})}{\partial \mathbf{u}}\right|_{\mathbf{x}_0,\mathbf{u}_0} \quad (58)$$

$$\mathrm{C} = \left.\frac{\partial \mathbf{g}(\mathbf{x},\mathbf{u})}{\partial \mathbf{x}}\right|_{\mathbf{x}_0,\mathbf{u}_0} \quad (59)$$

$$\mathrm{D} = \left.\frac{\partial \mathbf{g}(\mathbf{x},\mathbf{u})}{\partial \mathbf{u}}\right|_{\mathbf{x}_0,\mathbf{u}_0} \quad (60)$$

If matrix D is nonsingular, finally we have

$$\frac{d\mathbf{\Delta x}}{dt} = (\mathbf{A} - \mathbf{B}\mathbf{D}^{-1}\mathbf{C})\mathbf{\Delta x} = \mathbf{\Lambda \Delta x} \quad (61)$$

The eigenvalues and eigenvectors of the state matrix can reflect the stability of the system at the operating point and the characteristics of the oscillation (Kundur, 1994).
According to the established 7th order DFIG model described in Section 4, the state variables are I_{dr}, I_{drref}, $\hat{U}_{dr}$, $\hat{U}_{qr}$, I_{qr}, I_{qrref} and s, respectively, and the algebraic variables are U_{ds}, U_{qs}, I_{ds}, I_{qs}, respectively. In small signal stability analysis, when the wind farm is integrated into the power grid, these algebraic variables mentioned above in d-q coordinate system need to be transformed to the synchronous rotating coordinate system (x-y coordinate system), i.e. these algebraic variables will be subject to

$$\mathbf{U}_{dq} = \boldsymbol{T}\mathbf{U}_{xy} \quad (62)$$

$$\mathbf{I}_{dq} = \boldsymbol{T}\mathbf{I}_{xy} \quad (63)$$

Where T is transformation matrix, $\boldsymbol{T} = \begin{bmatrix} \sin\theta & -\cos\theta \\ \cos\theta & \sin\theta \end{bmatrix}$; θ denotes the phase angle of generator terminal voltage U_t, which can be obtained by the steady state power flow solutions.
Accordingly, we can obtain

$$\frac{d\Delta\mathbf{x}}{dt} = \boldsymbol{A}_{WF}\Delta\mathbf{x} + \boldsymbol{B}_{WF}\Delta\mathbf{U}_{xy} \quad (64)$$

$$\Delta\mathbf{I}_{xy} = \boldsymbol{C}_{WF}\Delta\mathbf{x} + \boldsymbol{D}_{WF}\Delta\mathbf{U}_{xy} \quad (65)$$

Where $\boldsymbol{A}_{WF}$, $\boldsymbol{B}_{WF}$, $\boldsymbol{C}_{WF}$ and $\boldsymbol{D}_{WF}$ are expressed as follows

$$
\begin{array}{c}
\begin{array}{ccccccc}
\quad \Delta I_{dr} & \Delta I_{drref} & \Delta \hat{U}_{dr} & \Delta I_{qr} & \Delta I_{qrref} & \Delta \hat{U}_{qr} & \Delta s
\end{array}\\
\boldsymbol{A}_{WF} = \begin{bmatrix}
-\frac{\omega_s}{T_r'} & & -\frac{\omega_s}{X_r'} & & & & \\
K_1 U_t \frac{X_m}{X_s}(\frac{\omega_s}{T_r'} - \frac{1}{T_1}) & & K_1 U_t \frac{X_m}{X_s}\frac{\omega_s}{X_r'} & & & & \\
-K_2(\frac{\omega_s}{T_r'} - \frac{1}{T_2}) & -\frac{K_2}{T_2} & -K_2 \frac{\omega_s}{T_2} & & & & \\
 & & & -\frac{\omega_s}{T_r'} & & -\frac{\omega_s}{X_r'} & \\
 & & & K_1 U_t \frac{X_m}{X_s}(\frac{\omega_s}{T_r'} - \frac{1}{T_1}) & & K_1 U_t \frac{X_m}{X_s}\frac{\omega_s}{X_r'} & \\
 & & & -K_2(\frac{\omega_s}{T_r'} - \frac{1}{T_2}) & -\frac{K_2}{T_2} & -K_2 \frac{\omega_s}{T_2} & \\
 & & & -U_t \frac{X_m}{2HX_s} & & &
\end{bmatrix}
\end{array}
\tag{66}
$$

$$
\begin{array}{c}
\begin{array}{cc}
U_{ds} & U_{qs}
\end{array}\\
\boldsymbol{B}_{WF} = \begin{bmatrix}
0 & 0 \\
0 & K_1 I_{dr} \frac{X_m}{X_s}(\frac{\omega_s}{T_r'} - \frac{1}{T_1}) + K_1 \hat{U}_{dr} \frac{X_m}{X_s}\frac{\omega_s}{X_r'} - 2\frac{K_1 U_t}{T_1 X_s} \\
0 & 0 \\
0 & 0 \\
0 & K_1 I_{qr} \frac{X_m}{X_s}(\frac{\omega_s}{T_r'} - \frac{1}{T_1}) + K_1 \hat{U}_{qr} \frac{X_m}{X_s}\frac{\omega_s}{X_r'} \\
0 & 0 \\
0 & -\frac{X_m}{X_s} I_{qr}
\end{bmatrix} \boldsymbol{T}
\end{array}
\tag{67}
$$

$$
\boldsymbol{C}_{WF} = \boldsymbol{T}^{-1} \begin{bmatrix}
-\frac{X_s}{X_m} & 0 & 0 & 0 & 0 & 0 & 0 \\
0 & 0 & 0 & -\frac{X_s}{X_m} & 0 & 0 & 0
\end{bmatrix}
\tag{68}
$$

$$
\boldsymbol{D}_{WF} = \boldsymbol{T}^{-1} \begin{bmatrix}
0 & -\frac{1}{X_s} \\
0 & 0
\end{bmatrix} \boldsymbol{T}
\tag{69}
$$

Here, each generator in a power system to be studied can be represented as the aforementioned form in accordance with the dynamic model of itself. For a power system consisting of n generators (including wind farm) with m state variables, by eliminating $\Delta\mathbf{I}_{xy}$, we can get the Jacobian matrices of the whole system **A**, **B**, **C** and **D** (Wang et al., 2008)as given in following

$$\mathbf{A} = \left[\mathbf{A}_G\right]_{m\times m} \tag{70}$$

$$\mathbf{B} = \left[\mathbf{B}_G \quad 0\right]_{m\times 2N} \tag{71}$$

$$\mathbf{C} = \begin{bmatrix} -\mathbf{C}_G \\ 0 \end{bmatrix}_{2N\times m} \tag{72}$$

$$\mathbf{D} = \begin{bmatrix} \mathbf{Y}_{GG} - \mathbf{D}_G & \mathbf{Y}_{GL} \\ \mathbf{Y}_{LG} & \mathbf{Y}_{LL} \end{bmatrix}_{2N\times 2N} \tag{73}$$

Where

$$A_G = \begin{bmatrix} A_1 & & & & \\ & \ddots & & & \\ & & A_{WF} & & \\ & & & \ddots & \\ & & & & A_n \end{bmatrix}_{m\times m} \tag{74}$$

$$B_G = \begin{bmatrix} B_1 & & & & \\ & \ddots & & & \\ & & B_{WF} & & \\ & & & \ddots & \\ & & & & B_n \end{bmatrix}_{m\times 2n} \tag{75}$$

$$C_G = \begin{bmatrix} C_1 & & & & \\ & \ddots & & & \\ & & C_{WF} & & \\ & & & \ddots & \\ & & & & C_n \end{bmatrix}_{2n\times m} \tag{76}$$

$$D_G = \begin{bmatrix} D_1 & & & & \\ & \ddots & & & \\ & & D_{WF} & & \\ & & & \ddots & \\ & & & & D_n \end{bmatrix}_{2n\times 2n} \tag{77}$$

Where, $\mathbf{Y}_{GG}$ and $\mathbf{Y}_{LL}$ are the self-admittance matrices of generator nodes and non-generator nodes; $\mathbf{Y}_{LG}$ and $\mathbf{Y}_{GL}$ are the mutual admittance matrices between them.

Finally, the corresponding state matrix can be given in following:

$$\mathbf{\Lambda} = \mathbf{A} - \mathbf{B}\mathbf{D}^{-1}\mathbf{C} \tag{78}$$

7. Probabilistic small signal stability analysis with wind farm

7.1 Principle of Monte Carlo simulation

Monte Carlo method is a class of computational algorithms that rely on repeated random sampling to compute their results. In uncertainty analysis, the relationship between the dependent variable and independent variable can be described as

$$\mathbf{Z} = \mathbf{h}(\mathbf{X}) \tag{79}$$

where $\mathbf{X} = [x_1, x_2, \dots, x_m]$ is the vector of the independent variables and $\mathbf{Z}=[z_1,z_2, \dots, z_n]$ is the vector of the dependent variable. $\mathbf{h}=[h_1(\mathbf{X}), h_2(\mathbf{X}), \dots, h_m(\mathbf{X})]$ represents the function relationship between input variable and output variable. In general, if **h** is very complex, it is hard to solve the probability distribution of **Z** applying analytic way. In this situation, the Monte Carlo methods are employed to calculate the discrete frequency distribution which approximately simulates its probability distribution. The essence of the uncertainty analysis is to estimate the statistic properties of **Z** based on the statistic properties of **X** and the function **h** (Fishman, 1996). The most important statistic property in uncertainty analysis is the probability distribution, which is always described by the probability density function (PDF). Probability density function describes the probability density of a variable at a given value (Fishman, 1996). Therefore, the main objective of uncertainty analysis is to estimate the PDF of the dependent variable on the basis of the PDF of independent variable and their relationship function. Monte Carlo simulation is a repetitive procedure: (1) The random independent variable **X** is generated based on its PDF; (2) According to the relationship function **h**, the vector **Z** can be calculated; (3) Repeat (1) and (2), the PDF of the dependent variable **Z** can be estimated when the sample size (the number of repetition) is large enough. The justification of Monte Carlo simulation comes from the following two basic theorems of statistics: (i) The Weak Law of Large Numbers and (ii) The Central Limit Theorem. Based on the above two theorems, it can be proved that with increasing of sample size, the PDF of the dependent variable obtained by Monte Carlo simulation will approach to that of the population.

7.2 Probabilistic small signal stability incorporating wind farm

The flow chart of the Monte Carlo simulation technique for power system small signal stability analysis with consideration of wind generation intermittence is given in Fig. 6.

It is well known that the uncertainty of wind generation is due to the uncertainty of wind speed, so we begin with the probability distribution of the wind speed. Fig. 7 shows a Weibull distribution function of wind speed with k = 2 and c =10. When a random wind speed is generated, the mechanical power output extracted from the wind can be calculated via a king of wind turbine model usually given by functions approximation. If the wind speed V_m is less than the cut-in speed $V_{cut\text{-}in}$ or is larger than the cut-off speed $V_{cut\text{-}off}$, the wind farm will be tripped. If the current wind speed belongs to the speed range from cut-in to cut-off, the wind farm will be kept connected to the grid in power flow calculation and small signal stability analysis. The process is repeated until the pre-set sample size N is reached. Finally, the probabilistic-statistical analysis can be conducted based on the results from different wind speed conditions mentioned above to reveal the impact of wind generation intermittence on power system small signal stability.

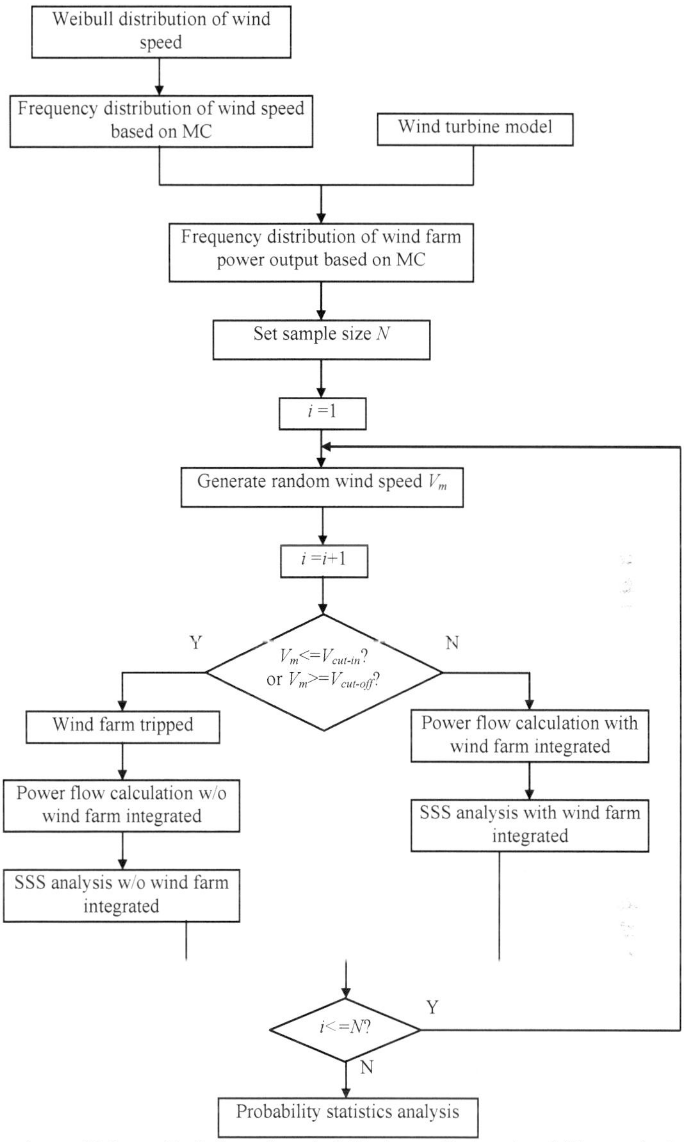

Fig. 6. Flow chart of Monte Carlo based probabilistic small signal stability analysis incorporating wind farm

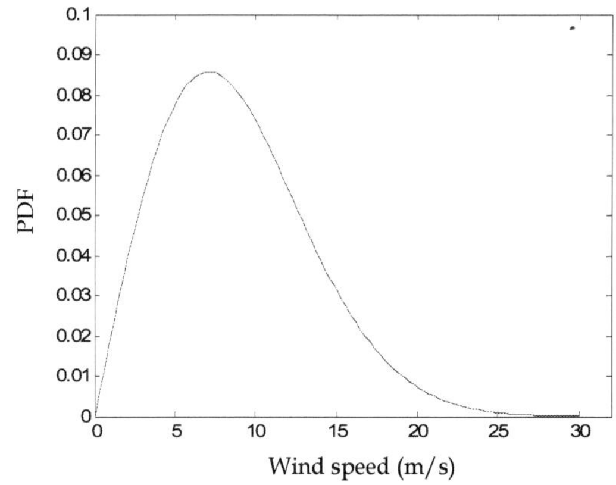

Fig. 7. Weibull distribution with $k = 2$ and $c = 10$

8. Application example

The IEEE New England (10-generator-39-bus) system was employed as benchmark to test the proposed model and method. The single line diagram of the test system is given in Fig. 8. In this system, the classical generator model is applied to the synchronous generator G2. The 4th order generator model with a simplified 3rd order exciter model is applied to the remaining 9 synchronous generators. It should be noticed that there is no any power system stabilizer considered in the test system. All simulations were implemented on the MATLAB™ environment.

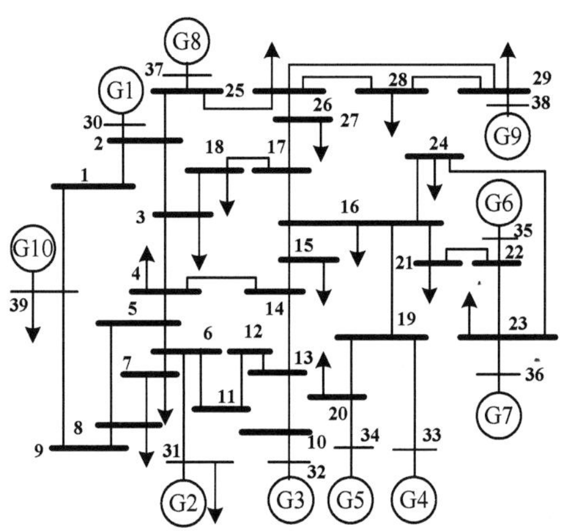

Fig. 8. Single line diagram of IEEE New England test power system

A wind farm with 200×2MW DFIGs is integrated into the non-generator buses, i.e. bus1-bus29. The corresponding parameters of wind turbine and DFIG are given in Table 1.

According to the procedure given in Fig. 6, the frequency distribution of wind speed by applying Monte Carlo method to the Weibull probability distribution of wind speed can be calculated as depicted in Fig. 9. The sample size is set to be 8000 during simulation.

Parameters	Values
ρ	1.2235 kg/m^3
R	45 m
C_p	0.473
$V_{cut\text{-}in}$	3m/s
$V_{cut\text{-}off}$	25m/s
V_{rated}	10.28m/s
R_s	0.00488
X_{ls}	0.09241
X_{lr}	0.09955
X_m	3.95279
R_r	0.00549
H	3.5
K_1	0.1406
T_1	0.0133
K_2	0.5491
T_2	0.0096

Table 1. Parameters of wind turbine and DFIG with 2 MW capacity

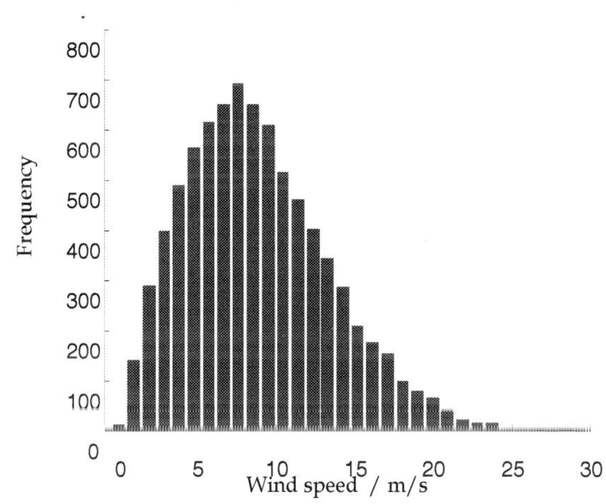

Fig. 9. Frequency distribution of wind speed

Next, in accordance Eq. (1) and the frequency distribution of wind speed as shown in Fig.9, for the wind farm with 200*2MW capacity, the probability distribution of wind farm power output can be finally obtained as shown in Fig. 10. From Fig.10, there exist two concentrations of probability masses in the distribution: one corresponds to the value of zero, in which the wind farm is cut off; the other corresponds to the value of 400MW, in which the rated power output is generated by the wind farm.

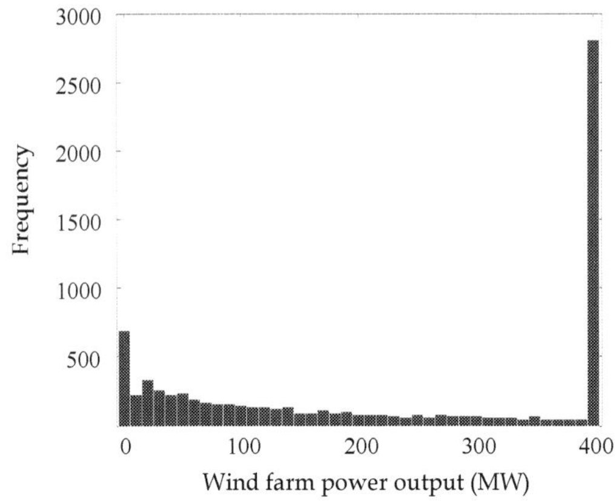

Fig. 10. Probability distribution of wind farm power output

Fig. 11 shows the frequency distribution of the real part of eigenvalues when the wind farm is connected to bus20.

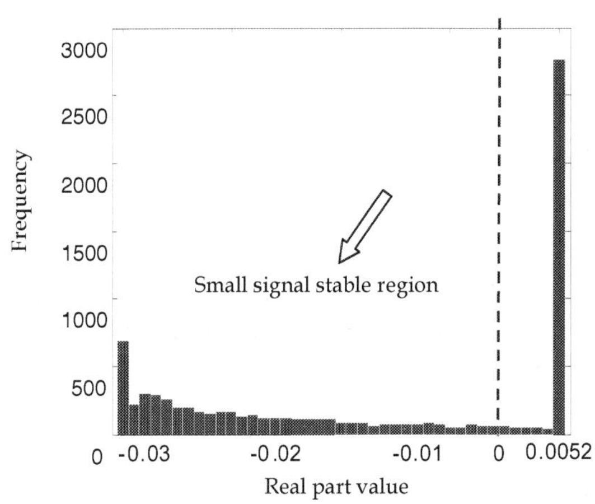

Fig. 11. Probability distribution of real part of eigenvalues with wind farm integration into bus20

According to the statistical analysis based on Fig. 11, we found that there is a probability of roughly 39.1% (3128 out of 8000 in simulation) that the real part of the eigenvalue will be positive, in which situation the system is small signal unstable. Therefore we can conclude that the stability probability of the test system is 60.9% in current operating condition. Furthermore, there exist two concentrations of probability masses in the distribution: the left one corresponds to the situation that the wind farm is cut off; the right one corresponds to situation that wind turbine generates rated power.
Under the same wind speed condition, the wind farm is connected to the bus1-29, respectively. The corresponding results are given in Table 2.

Bus	No. of wind generator	Stable Probability
20	200	60.9%
Others	200	100%

Table 2. Small signal stable probability with different wind farm integration position

Electro-mechanical oscillation mode can be picked out according to the electro-mechanical relative coefficient or the frequency of oscillation, i.e. $\rho>1$ or $0.1<f<2.5$Hz (Wang et al., 2008). The mean and standard error of the mode properties: frequencies, Electro-mechanical relevant ratio, damping ratio, and the participating factors are given in Table 3. We found that the 9th EM oscillation mode is unstable with a probability of 39.1%, and it is actually the pair of eigenvalue that determines the small signal stability probability of the whole system.
In summary, according to the simulation results discussed above, we can conclude that the deterministic small signal stability analysis can be considered as a special case study in the probabilistic small signal stability analysis. Especially, the probabilistic small signal stability based on the Monte Carlo method can evaluate the test power system more objectively and accurately.

9. Conclusion

This chapter addresses the impact of intermittent wind generation on power system small signal stability. Firstly, the well-known Weibull probability distribution is employed to reveal wind speed uncertainty. According to the Weibull distribution of wind speed, the Monte Carlo simulation technique based probabilistic small signal stability analysis is applied to solve the probability distributions of wind farm power output and the eigenvalues of the state matrix. Finally, the IEEE New England test power system is studies as benchmark to demonstrate the effectiveness and validity of the propose model and method. Based on the numerical simulation results, we can determine the instability probability of the power system with the uncertainty and randomness of wind power consideration. And from viewpoint of small signal stability, the most suitable integration position for wind farm can be determined as well.

EM mode	Frequency		Damping ratio		EM relevant ratio		Most relevant generator	Stability
	Mean	Std	Mean	Std	Mean	Std		
1	1.5021	0.0000	0.0506	0.0002	22.9116	0.1119	G8	100%
2	1.4816	0.0003	0.0613	0.0002	36.1465	3.9023	G7	100%
3	1.4569	0.0006	0.0642	0.0010	10.2773	0.3156	G4	100%
4	1.2794	0.0001	0.0363	0.0005	29.5096	0.2265	G8	100%
5	1.2642	0.0013	0.0133	0.0026	91.3249	25.3277	G2	100%
6	1.1370	0.0024	0.0375	0.0001	32.0537	0.0459	G6	100%
7	1.0394	0.0040	0.0084	0.0027	35.6480	1.9408	G9	100%
8	0.9817	0.0012	0.0066	0.0019	48.0187	8.0547	G6	100%
9	0.6554	0.0020	0.0030	0.0033	52.8691	4.1706	G10	60.9%

Table 3. Properties of EM oscillation modes with wind farm integration into bus20

10. References

Ackermann, T. (2005). *Wind Power in Power Systems,* Wiley, ISBN 0-470-85508-8, Chichester

Akhmatov, V. (2002). Variable-speed wind turbine with doubly-fed induction generators part i: modelling in dynamic simulation tools. *Wind Engineering,* Vol. 26, No. 2, 85-108, ISSN 0309-524X

de Alegría, IM.; Andreu, J.; Martín, JL.; Ibañez, P.; Villate, JL. & Camblong, H. (2007). Connection requirements for wind farms: A survey on technical requierements and regulation, *Renewable and Sustainable Energy Reviews*, Vol.11; No.8, 1858 - 1872, ISSN 1364-0321

Feijóo, A.; Cidrás, J. & Carrillo, C. (2000). A third order model for the doubly-fed induction machine, *Electric Power Systems Research,* Vol.56, No. 2, 121-127, ISSN 0378-7796

Fishman, G. (1996). *Monte Carlo: Concepts, Algorithms, and Applications,* Springer-Verlag, ISBN 978-0387-9452-79, New York

Joselin, H. G.; Iniyan, S.; Sreevalsan, E. & Rajapandian, S. (2007). A review of wind energy technologies, *Renewable and Sustainable Energy Reviews,* Vol.11, No.6, 1117–1145, ISSN 1364-0321

Kundur, P. (1994). *Power System Stability and Control,* McGraw-Hill, ISBN 7-5083-0817-4, New York

Mei, F. & Pal, B. C. (2005). Modelling and small-signal analysis of a grid connected doubly-fed induction generator, *IEEE Power Engineering Society General Meeting*, pp. 2101–2108, ISBN 0-7803-9157-8, San Francisco, June 2005, IEEE, New Jersey

Mendonca, A. & Lopes, J. A. P. (2005). Impact of large scale wind power integration on small signal stability, *Future Power System International Conference*, pp. 1-5, ISBN 90-78205-02-4, Amsterdam, Nov. 2005, IEEE, New Jersey

Rouco, L. & Zamora, J. L. (2006). Dynamic patterns and model order reduction in small-signal models of doubly fed induction generators for wind power applications, *IEEE Power Engineering Society General Meeting*, pp. 1–8, ISBN 1-4244-0493-2, Montreal, July 2006, IEEE, New Jersey

Rueda, J. L.; Colome, G. D. & Erlich, I. (2009). Assessment and enhancement of small signal stability considering uncertainties. *IEEE transaction on Power Systems*, Vol. 24, No. 1, 198-207, ISSN 0885-8950

Slootweg, J. G.; Polinder, H.; & Kling, W. L. (2001). Dynamic modelling of a wind turbine with doubly fed induction generator, *IEEE Power Engineering Society Summer Meeting*, pp. 644–649, ISBN 0-7803-7173-9, Vancouver, July 2001, IEEE, New Jersey

Tapia, G.; Tapia, A. & Ostolaza, J. X. (2006). Two alternative modeling approaches for the evaluation of wind farm active and reactive power performances. *IEEE Transaction on Energy Conversion*, Vol. 21, No. 4, 909 – 920, ISSN 0885-8969

Tapia, A.; Tapia, G. & Ostolaza, J. X. & Saenz, J. X. (2003). Modeling and control of a wind turbine driven doubly fed induction generator. *IEEE Transaction on Energy Conversion*, Vol. 12, No. 2, 194 – 204, ISSN 0885-8969

Tsourakisa, G.; Nomikosb, B. M. & Vournasa, C. D. (2009). Effect of wind parks with doubly fed asynchronous generators on small-signal stability. *Electric Power Systems Research*, Vol.79, No.1, 190–200, ISSN 0378-7796

Wang, C.; Shi, L.; Wang, L. & Ni, Y. (2008). Modelling analysis in power system small signal stability with grid-connected wind farms of DFIG Type, *Wind Engineering*, Vol.32, No.3, 243–264, ISSN 0309-524X

Wang, K. W.; Chung, C. Y.; Tse, C. T. & Tsang K. M. (2001). Probabilistic eigenvalue sensitivity indices for robust PSS site selection. *IEE Proceedings- Generation, Transmission and Distribution*, Vol. 148, No. 4, 603-609, ISSN 1350-2360

Wang, K. W.; Chung, C. Y.; Tse, C. T. & Tsang, K. M. (2000). Improved probabilistic method for power system dynamic stability studies, *IEE Proceedings- Generation, Transmission and Distribution*, Vol.147, No.1, 37-43, ISSN 1350-2360

Wang, X. F.; Song Y. H. & Irving, M. (2008). *Modern Power System Analysis*, Springer, ISBN 13-9780-3877-2852-0, New York

Wu, F.; Zhang, X. P.; Godfrey, K. & Ju, P. (2006). Modeling and control of wind turbine with doubly fed induction generator, *IEEE Power Systems Conference and Exposition*, pp. 1404–1409, ISBN 1-4244-0177-1, Atlanta, Nov. 2006, IEEE, New Jersey

Xu, Z.; Dong, Z. Y. & Zhang, P. (2005). Probabilistic small signal analysis using Monte Carlo simulation, *IEEE Power Engineering Society General Meeting*, pp. 1658–1664, ISBN 0-7803-9157-8, San Francisco, June 2005, IEEE, New Jersey

Part 5

Spin-off Products of Offshore Wind Farms

9

The Potential for Habitat Creation around Offshore Wind Farms

Jennifer C. Wilson
AMEC
UK

1. Introduction

The growth of offshore renewable energy generation is the biggest expansion of development in the marine environment in recent years, with offshore wind farms at the forefront of this. Due to its favourable wind resource, Europe in particular is rapidly expanding its portfolio of offshore wind energy generation; however, the rest of the world is also beginning to take advantage of this natural resource. This is due, in part, to the fact that Europe, especially the north-west region, has ideal conditions for development, due to the high offshore wind levels, and the fact that its coasts slope gently away from the land. This means that water depths increase relatively slowly in most areas, making conditions highly suitable for offshore construction (Ackermann and Soder, 2002).

In addition to this, the offshore wind environment is much more reliable than onshore wind, as it is less turbulent, and has a higher energy density. This is due to the convection caused by the differential heating and cooling of the land and sea over the daily cycle, making the offshore zone generally windier. Further offshore, the lack of surface roughness adds to average wind speeds, further increasing energy efficiency. It is estimated that an offshore wind farm can generate around 50% more electricity than can be generated from an equivalent sized land-based development (Linley *et al.*, 2007).

In the UK, the development of offshore wind energy generation has been undertaken in a series of Rounds. In April 2001, following a detailed consultation and application process, eighteen 'Round 1' sites were announced, with a maximum of 30 turbines (BWEA, 2005). Whilst these projects were in the planning stages, further consultation was undertaken, discussing topics which would be critical to future development, such as the consents process, legal frameworks and the electrical infrastructure required for future projects. Three Strategic Areas in UK waters were identified, with fifteen projects being granted permission to submit formal applications under 'Round 2'. In January 2010, a further nine zones were allocated to developers through a competitive application process, under 'Round 3'. On top of these, there have also been Round 2 extensions granted for certain projects, and a number of sites granted exclusivity agreements to apply for development in Scottish Territoral Waters.

In 2008, the UK overtook Denmark to become the world-leader in generating energy from offshore wind (Jha, 2008). With current UK emphasis on the construction of Round 2 projects, and the early development phases of Round 2 extensions, Round 3 and Scottish Territorial Waters projects, there is the potential for thousands more turbines to be installed in the waters around the UK, with expansion also predicted for many other countries world-wide, as technology develops.

As with the expansion of any relatively 'young' industry, there are concerns over the potential for environmental impacts resulting from offshore wind farms, including damage to the seabed from the installation of the turbines, and from the temporary placement of jack-up vessels, generally used in the construction of offshore wind farms.

Of the four phases of an offshore wind farm (exploration, construction, operation and decommissioning), it is generally considered that for marine life, the construction period has the greatest potential for causing impacts. It is inevitable that the installation of a foundation and tower (currently up to around 6m in diameter) will cause the removal of an area of the receiving seabed as available habitat for infaunal and epifaunal species (animals living in and on the seabed). For immobile species, this can also result in mortality, either through the impact itself, or the noise from the piling hammer, if the foundations are to be driven into the seabed.

The potential impacts arising from the various phases of the project are illustrated in Figure 1, taken from Elliott (2002). These diagrams are not exhaustive, but give a good indication of the intricacies of the impacts which may be caused by the installation of an offshore wind farm.

It should be noted that the impacts demonstrated in Figure 1 are heavily weighted towards marine life, i.e. benthic fauna, fish and marine mammals. Other impacts include potential impacts on bird populations (more complex than demonstrated here), as well as socio-economic issues, such as changes in levels of tourism in an area, and possibilities for job opportunities.

As with any developing industry, focus has been on the potentially negative impacts on the environment, so that these can be reduced, and where possible, eliminated. In this light, one element of offshore wind farms which has yet to be fully investigated and acknowledged at a wider level, is the potential for the submerged towers and foundations of the turbines to act as artificial reefs, with the capacity to increase the abundance and diversity of species and habitats within the receiving environment.

This chapter aims to review the current body of work in this area, looking at the following areas:

- *Artificial reefs in the marine environment*: Almost any structure in the marine environment has the potential to be colonised by marine life, thereby acting as an artificial reef, whether intended for the purpose or not. As the issue of marine conservation grows in importance, the installation of structures specifically for the purpose of enhancing abundance and diversity has increased, along with the body of work into rates of colonisation and suitability of various materials for the purpose.
- *Current evidence for offshore wind farms acting as artificial reefs*: Although there is still a relatively low number of fully-constructed and operational offshore wind farms around the world, there are a number of studies which have looked at the way in which marine life interacts with the turbines and their associated scour protection, where deployed. This includes post-construction surveys, as required in the conditions of consent, as well as scientific studies, looking to further knowledge on potential impacts and benefits.
- *Potential habitat enhancement by offshore wind farms*: Once more is understood about the interactions between offshore wind turbines, any associated scour protection, and the marine environment into which they are installed, it may be feasible to adapt design or deployment methods in order to maximise the benefit to the environment.

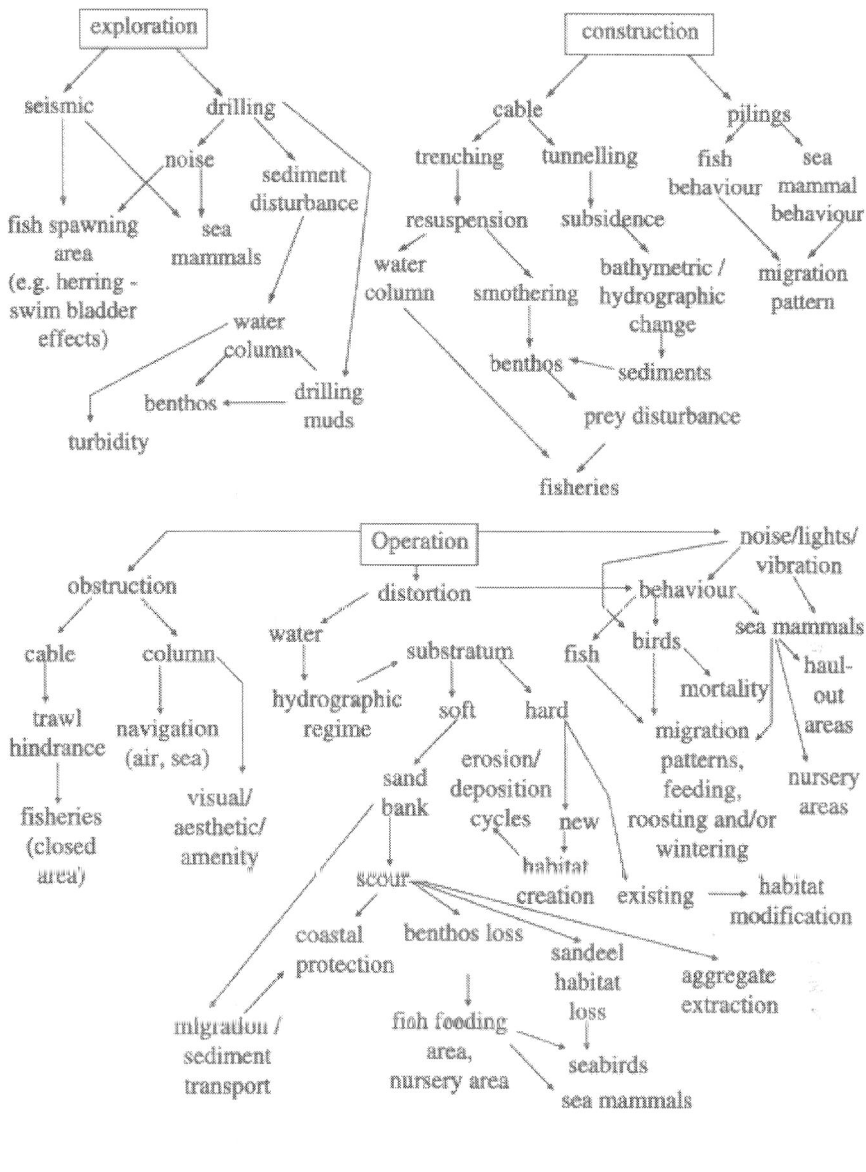

Fig. 1. Environmental impacts of offshore wind farms during pre-installation exploration, construction (similar effects are likely to occur during decommissioning) and operation (adapted from Elliott, 2002).

Finally, areas of future study requirements will also be addressed. As with almost all areas of study, the potential for habitat creation around offshore wind farms will benefit greatly from additional work in the field. With more developments planned, more surveys around operational wind farms will determine the significance of habitats created around the turbines, as well as assisting in possible modifications to future designs, construction plans or survey methodologies. Further work on artificial reef deployment in general will also add to understanding.

2. Artificial reefs in the marine environment

Already, there exists a large body of both anecdotal and scientific data on the benefits of artificial reefs - both intentionally created and otherwise - to marine life. Anyone who has seen a pier support or harbour wall will know how rapidly colonisation of any introduced surface into the environment can occur, with the initial populations soon attracting more individuals and species to the newly-developed community. A number of studies have focused on the sequence of colonisation, how rapidly it occurs, and what benefits it can have to the surrounding environment.

For many years, there has been anecdotal evidence of oil rig workers fishing from platforms, reporting high numbers of large fish, suggesting that the fish were using the reefs as shelter in an otherwise featureless ocean environment. Lokkeborg *et al.* (2002) conducted a study around two North Sea platforms, one partly decommissioned and one still operational, using gill nets. It was found that catch rates increased rapidly close to the platforms, indicating a distinct increase in fish abundance (a linear relationship between catch rate and fish abundance was assumed in the study). Similar results have also been found around oil rigs in the Gulf of Mexico. It is thought that shelter from prevailing currents, lower risk of predation and higher prey densities all contribute to the tendency for fish to aggregate around oil rigs (Lokkeborg *et al.*, 2002).

Several projects around the world have taken advantage of this function of oil rigs, including the Louisiana Artificial Reef Programme, established in 1986 to take advantage of the obsolete oil and gas platforms which had been shown to be important habitats for the region's fish populations. It was recognised that to remove the platforms once decommissioned would be to remove potentially valuable habitat from the environment, despite regulations that platforms be removed a year after the end of production (Louisiana Department of Wildlife and Fisheries, 2005). Since the installation of the first platform in the region in 1947, it was noted that fishermen of Louisiana and neighbouring states had recognised the value of the surrounding waters as fishing grounds, with the structures being the destination of over 75% of recreational fishing trips departing from Louisiana (Wilson *et al.*, 1987). When it became apparent that the majority of the rigs would be removed on decommissioning, the project was launched in order to save the habitat and resulting fish populations. The programme followed similar ventures in South Carolina, Alabama and Florida (with one of the first documented artificial reefs being initiated by a private individual in the 1800s in South Carolina), as well as in other countries. Following a large-scale consultation with key user groups, including local fishermen, who were hoping to benefit most from the programme, several sites were selected for the structures to be located. It has been estimated that a single 4-pile platform jacket (standard construction for underwater support for a platform) can provide between 2 and 3 acres of habitat (Bureau of Ocean Energy Management, Regulation and Enforcement, 2010), a valuable addition to a flat, plain environment, dominated by mud, clay and sand, with very little natural rock bottom or reef habitat.

A number of research programmes have followed the development of the rig structures as artificial reefs, including those undertaken by the Minerals Management Service's own divers, who recorded plant and invertebrate colonisation within only a couple of weeks of installation. Within a year of first installation (as an operational rig), the rig can be completed covered, and already forming the base of a highly complex food chain. Researchers found that fish densities could be up to 50 times greater around the sunken platforms, with each former rig serving as habitat for between 10 and 20 thousand individual fish, many of commercial or recreational importance for the region (Bureau of Ocean Energy Management, Regulation and Enforcement, 2010). Although not all rigs are utilised by the rigs-to-reef programme, every one which is has the potential to bring about large benefits for the surrounding marine environment. They have also been found to be of benefit economically, with recreational charter boats, fishermen, and diving operators all listing the rigs as amongst their most popular destination for recreational fishermen and divers, both keen to take advantage of the rich biodiversity the rigs create. The programme is so successful that in 2002 it was recognised as such, with the main leaders of the project receiving special citation at the Offshore Technology Conference, Houston.

Other structures have been introduced to the marine environment with the direct aim of enhancing the populations in the surrounding area, as well as bringing possible economic benefits through the attraction of human visitors. Large-scale examples of this are ships such as HMS Scylla, off the south coast of England in 2004, and more recently, in 2009, HMAS Canberra off Australia. These ships are often scuttled with the deliberate aim of creating habitat for both marine and human life, namely in the shape of SCUBA divers. For the Scylla, the main purpose for the sinking was the creation of a purpose-built, safe dive site, bringing in high-spending divers to the area. However, it has also presented local scientists with an opportunity to study colonisation of an underwater structure from just days after entering the water. Surveys showed that after only 10 days, fish had started to use the area, followed by tube worms, barnacles, hydroids etc, and wandering species such as crabs. After ten weeks, there was significant variety of life on the wreck (Hiscock, 2009). By the end of the first year of survey work, 53 species had been recorded on or in the Scylla, with the sequence of colonisation and loss of species being traceable through regular study. In March 2009, it was reported that 258 species had been recorded on the Scylla (Hiscock, 2009), and although a number of 'expected' species were yet to be noted on the wreck, and some species were not found in the abundances expected after five years, it is still a significant increase in abundance and diversity for the immediate area. It has also become a major diving attraction for the area.

One of the key issues with artificial reefs is whether the installation is actually producing its own life, and thereby contributing to the surrounding community, or simply attracting life away from nearby habitats, and therefore perhaps actually having a negative effect, by 'thinning out' local populations. A number of studies have investigated this in relation to fish or motile invertebrate communities, but there is little work done on benthic communities (Perkol-Finkel and Benayahu, 2007). Part of the difficulty in determining whether artificial reefs simply divert propagules from their natural destinations, or attract those which would otherwise be lost, is due to the difficulty of following larval movements in the ocean, despite many advances in this field. In their 2007 study, Perkol-Finkel and Benayahu undertook experiments in the Red Sea, using settlement plates to determine any differences between artificial and natural reefs. It was found that recruitment of fouling invertebrates and corals clearly differed between the artificial and natural reef areas, both in species composition and abundance. It was suggested therefore, that the majority of the organisms which colonised the artificial reef area were not derived from adjacent natural reefs, and so in all likelihood, would not have been recruited to the area were it not for the artificial reef being present (Perkol-Finkel and Benayahu, 2007). It is therefore noted that artificial reefs are able to increase the species diversity of an area, perhaps through the introduction of different conditions, habitat types and available niches, to those already available naturally.

3. Current evidence for offshore wind farms as artificial reefs

Due in part to the youthful nature of the offshore wind industry, there are still relatively few fully comprehensive studies into the influence of turbine arrays on fish and benthic populations, other than the monitoring requirements set out in the consent conditions. However, where datasets do exist, it is suggested that offshore wind farms are demonstrating benefits for such populations.

The effect on commercial stocks, such as lobster and crab, are an obvious concern to those directly and indirectly involved in the exploitation of such stocks; therefore any impacts are key to the Environmental Impact Assessment (EIA) process. Observations made onboard a commercial potting vessel deploying gear within the operational Barrow Offshore Wind Farm, off the north west coast of England, eighteen months after construction was completed, found that catch rates for lobster were similar inside and outside of the wind farm boundary (Centrica, 2009). In addition to this, the number of undersized crabs taken within the wind farm was greater than the number found outside the boundary, suggesting that the wind farm site is acting as a haven for juvenile crabs. Initial thoughts that this may be due to lack of fishing effort, with the wind farm acting as an unofficial nature reserve, were discounted in the case of Barrow due to anecdotal evidence, which stated that potting had recommenced within the wind farm boundary a matter of weeks after construction was completed (Centrica, 2009).

A recent study by Langhamer and Wilhelmsson (2009) looked into the colonisation of wave power devices off the Swedish coast, with some of the foundations being perforated with holes at different heights and positions around the block foundations, to determine whether this would have a positive influence on colonisation. Surveys on the blocks were carried out by divers. Although fish populations in the area were generally relatively low, it was found that numbers were significantly higher around the foundations than in the control sites (sites of the same area, generally of sandy seabed, near to the foundations). Although the number of lobsters found was low, with individuals inhabiting crevices around the base of the foundation rather than the drilled holes, the foundations were found to have a positive effect on the number of edible crab, which increased around foundations with or without holes (Langhamer and Wilhelmsson, 2009).

At the Horns Rev Offshore Wind Farm, off the Danish coast, Forward (2005), found that in terms of benthic community structure, there was no significant difference between the wind farm site and a reference area. However, there was a substantial increase in the density of sand eels, rising by 300% within the operational wind farm in 2004, compared to a rise of only 20% at the reference site. This increase within the wind farm was mainly due to an increase in the number of juvenile sand eels, with the main reasons behind the increase thought to be reduced mortality through predation, and a reduction in mean particle size as a result of construction. In addition to this, eight new species were recorded within the wind farm site, compared to pre-construction surveys (Forward, 2005).

A number of studies have investigated the potential for offshore wind turbines to act as fish aggregating devices (FADs). FADs are not a modern phenomenon, and have been employed for centuries to concentrate marine fish and ease their capture, proving highly successful (Fayram and de Risi, 2007). In open-water areas, the catch-rates of some tuna species have been found to be 10-100 times greater near FADs, based on mark and recapture studies. This would clearly benefit local fish communities, of both commercial and non-commercial species, and where commercial stocks exist, would have the potential of enhancing such stocks for the local fishing industry. However, there is need for caution to be exercised here. It has been noted that in some situations, juveniles of some species are more associated with FADs than adult fish, thereby potentially resulting in the increased catch-rates of juveniles over adults, should these areas be fished (Fayram and de Risi, 2007).

This element would need further survey work before the true benefits for the fishing industry, if any, could be estimated.

Wilhelmsson *et al.* (2006) undertook research into the effects on fish populations at five wind farm sites in Sweden, and found that large communities of both demersal and pelagic fish populations developed around the turbines. It was noted that the presence of such populations may in fact lead to further enhancement of benthic communities around the base of the turbines, as a result of the deposition of organic material such as faecal matter, organic litter and dead organisms, all of which provide material for benthic organisms to feed on. In addition, it was reported that mussel beds were starting to develop in the areas adjacent to the wind turbines, possibly as a result of mussels being dislodged from their original attachment locations on the towers. A cyclical effect could develop here, as more benthic organisms means more food for fish, which increases the level of organic waste, thereby allowing further growth of benthic organisms, and so on.

The development of mussel populations on turbines could itself be of interest to the fishing community, as it was noted that previous studies have identified a link between mussel beds and increased fish numbers (Wilhelmsson *et al.*, 2006). Given that mussel growth is present on almost all turbine structures in the correct environmental conditions, this could be of particular interest.

The potential for the advantages of offshore wind farms acting as artificial reefs, and the ever-growing interest in the industry, means that there are frequently new research projects being designed to look into their capacity for colonisation and production.

The development of life around turbines is of key interest to the owners of the wind farms, as excessive build up of life can be damaging for the turbine. Surveying around the towers can also be specified as a condition of consent. For the operational Barrow Offshore Wind Farm, in the East Irish Sea, near Barrow-in-Furness, the surveying of colonisation of the monopile foundations and scour protection was required as part of the Food and Environment Protection Act (FEPA) licence granted for the project. The turbines had been installed in 2005, with the surveys being undertaken in 2008 (EMU, 2008a), consisting of video footage, still photography and sample collection by divers.

It was noted that on the four turbines surveyed, colonisation had taken place in a generally similar pattern, with a gradual change in community observed as depth increased. At the intertidal level on the turbines, there was found to be green algae, with barnacles slightly lower, giving way to increasingly dense populations of mussels moving down the tower. As depth increased, anemones increased in number, with mussels decreasing, with crabs and barnacles also being found. Around the base of the monopile was an area of coarse sediment, including shell fragments, pebbles and gravel (EMU, 2008a).

In general, the communities observed were typical of hard-surface communities, and commonly found in waters around the UK and Ireland. It was noted that no species of particular conservation interest or invasive / alien species had been found during the surveys. The results of the 2008 surveys were compared to initial survey work undertaken on six turbines, in 2006, around eight months after construction was completed. It was found that in general, the species found were similar between the two surveys, with abundances and densities increasing in the two years between surveys, as would be expected.

Further comparison was also made with surveys undertaken on the North Hoyle Offshore Wind Farm, in Liverpool Bay, completed one year after construction. Again, broadly similar communities were found to be developing on the turbine towers (EMU, 2008a). There was found to be minor variations in community structure; however, this is to be expected given the different locations, and therefore differing environmental influences.

Similar survey work has been undertaken on the Kentish Flats Offshore Wind Farm (EMU, 2008b), in the outer Thames Estuary, approximately three years after the installation of the turbines. In this survey, two turbines were assessed, and again, similar patterns of colonisation were found on each tower. Again, a change in community with depth was noted, with barnacles and mussels dominating the intertidal and infralittoral zones of the tower. As depth increased, mussels became scarcer, being replaced by anemones, with hydroids also becoming more prevalent. As with the other developments, at the base of the towers, shell fragments, pebbles and gravel dominated the seabed, with a number of crab species being found, as well as high numbers of starfish, unsurprising given the high densities of mussels, their key prey species (EMU, 2008b).

These studies show the capacity for colonisation within just a few months of the turbines being installed. Although in these cases, there has not been significant variation between turbines, or even wind farms, it is still a useful contribution to the productivity and ecological carrying capacity of the surrounding marine environment, with the potential to attract other species into the area looking for food sources, as the community continues to develop.

4. Potential habitat enhancement by offshore wind farms

As stated previously, the introduction of turbines and their associated scour protection has the capacity to increase the abundance and diversity of both species and habitats. The level of increase depends on the type of scour protection deployed, with the three main materials - boulders, gravel and synthetic sea-fronds - being included in a study which aimed to quantify the amount of habitat area created.

The need to deploy scour protection around the base of turbines depends on a number of factors, including seabed type, potential for seabed movement, and the design of the turbines themselves. Where used, as stated above, there are three main types of protection deployed, as illustrated in Figure 2.

Figure 2a illustrates the general scale of boulder or gravel protection around the base of a wind farm, for relative scales compared to average turbine dimensions. Although actual dimensions vary with specific turbine makes and models, and deeper water will bring about new designs and technologies, in general, projects currently under construction, or well-advanced in the planning process are in waters up to around 30m. Projects entering the planning process now, such as some in Scottish Territorial Waters, or as part of the large Round 3 zones, are in waters of 50m or more. The majority of turbines installed globally to date follow the same design as illustrated in Figure 2, the monopile design, with a single pile driven or drilled into the seabed, with the tower, nacelle and blades fitted on top. This is the foundation design which was used in the calculations by Wilson and Elliott (2009), the results of which are discussed below.

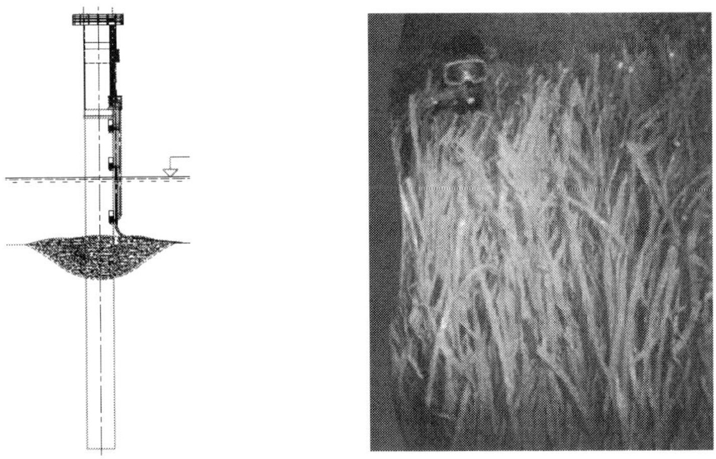

(a) (b)

Fig. 2. a) Approximate extent of rock / gravel protection around the base of a monopile wind turbine foundation; and b) Polypropylene frond mats around a foundation. Both taken from Linley *et al.* (2007).

From Figure 2b, it can also be seen that the synthetic frond mattresses can be relatively large in height, allowing plenty of shelter and protection for a wide variety of fish species.
Wilson and Elliott (2009) assessed the level of habitat lost and gained through the installation of a 4m diameter turbine, with an area of scour protection extending 10m from the base of the turbine. The results of the calculations from this study are shown in Table 1.

	Area (m^2)		
	Gravel Protection	Boulder Protection	Synthetic Sea-fronds
Seabed lost through turbine installation	452	452	452
Habitat created by scour protection	1102	1029	439.5
Net habitat loss / gain	650 (gain)	577 (gain)	-12.5 (loss)

Table 1. Habitat loss / gain due to the installation of an offshore wind turbine and associated scour protection. For these calculations, a turbine foundation diameter of 4m was assumed, with 10m of scour protection extending from the edge of the foundation. For gravel, a mean diameter of 5cm was assumed, with a 2m diameter for boulders.

From Table 1, it can be seen that for each turbine, there will be a gain in the surface area available for colonisation through the use of gravel or boulders as scour protection. For synthetic sea-fronds, although the values indicate a reduction in surface area, the change in habitats available should be noted. As offshore wind farms are generally located in relatively flat, biologically-sparse areas of seabed in order to reduce impacts on the seabed and associated organisms, the introduction of a sea grass type habitat will increase habitat diversity, thereby potentially still having ecological benefits for the area.

Each of the main scour protection materials has the potential to attract a distinct biological community, based on the type of habitats it can create, e.g. the level of shelter provided, or the lower organisms which are initially attracted to the structure.

Where gravel protection is deployed, the area will generally be inhabited by low numbers of robust polychaetes or bivalves, with occasional epibiota including echinoderms and crustaceans. According to the JNCC 2004/5 Comparative Tables, available through the JNCC website, other dominant species can include the parchment worm, which create extensive 'beds', within which can be found large populations of shrimps and small crab species, in turn providing food for species such as pipefish and seahorses, which are able to anchor themselves to the tubes by their tails (Anthoni, 2006).

Pipefish and seahorses are also amongst the species most likely to be found where synthetic sea-fronds have been used as scour protection, which, from anecdotal and photographic evidence, most closely mimics a sea grass bed once semi-buried by accumulating sediment. Sea grass beds are important habitats for fish, providing shelter from predation, nursery areas and refuges from larvae, as well as feeding grounds (Kopp *et al.*, 2007). Gobies also make up a large component of the sea grass community, with densities within sea grass beds reaching up to four times those in surrounding non-grassed areas (Pihl *et al.*, 2006).

Of the three scour protection materials, boulder protection has perhaps the greatest potential to enhance populations of commercially-fished species. If well designed, then lobster, edible crab and velvet swimming crab may be attracted, as well as reef fish such as wrasse and conger eels (Hiscock *et al.*, 2002), as the boulder protection will mimic rocky outcrops, which generally have higher levels of biodiversity and abundance than surrounding sandy seabed areas.

Due to the commercial status of lobster, a number of studies have been undertaken into how populations may be impacted / influenced. For example, one study looking into the settlement patterns of juvenile lobsters found that no lobsters were recorded settling onto sandy areas of seabed, compared to 19 lobsters/m^2 on large cobble and boulder covered areas (Linnane *et al.*, 2000). Work focusing on the colonisation of wave power foundations (Jensen *et al.*, 1994), suggested that the deployment of such structures into areas where lobster populations were habitat-limited could have the potential to enhance biomass production. It has been noted that shelter from predation may be a serious bottleneck for many species, lobster and crab included, therefore the deployment of wind and wave energy structures and associated boulder protection may increase production at a local scale (Langhamer and Wilhelmsson, 2009).

A major argument for the capacity for habitat creation around offshore turbines is the increased level of habitat diversity which is brought about through the introduction of a new habitat, whether it be rocky outcrop, gravel bed or sea grass patch. Diversity of available habitats is important in bringing about diversity in the number of species able to colonise and thrive in an area, and by mixing the various types of scour protection material within the same wind farm, it may be possible to bring about all three new habitat types, and the animals and plants which they attract.

There is also the potential for fin-fish species to benefit from the installation of turbine structures, with any of the associated scour protection materials deployed. As discussed above, the newly-created habitat will either attract in, or increase the productivity of, a wide range of species, including prey species for fin-fish. Increased productivity in the benthic community will, over time, enhance productivity all the way up the food chain, to the larger fish species, and potentially even marine mammals.

A further aspect of the habitat-creation benefits of offshore wind farms which must be considered is the deliberate targeting of scour protection and the materials deployed to directly benefit specific populations. On a simple level, this may involve using boulder protection in an area where there is an established lobster or crab fishery, in order to provide additional habitat, and improve productivity, as demonstrated by Linanane *et al.* (2000). By making deliberate attempts to increase the number of juveniles settling in an area, and ensuring the correct habitat type is available for adult lobsters, this has the relatively rare effect of encouraging both ecological and commercial benefits, in addition to the environmental gains of the renewable energy generated from the wind farm itself.

Taking this further, there are a number of specially-designed materials which could be easily adapted to be suitable for scour protection. One such example is the reef ball, designed and marketed by the Reef Ball Foundation, a non-profit environmental Non-Governmental Organisation (NGO), based in America. These structures come in a range of styles and sizes, designed to suit varying types of environment and seabed community. In general though, they are concrete domes, with a number of holes drilled into them at various levels and of various sizes, to provide a range of habitats for different species groups to utilise. Figure 3 shows the standard reef ball design. More complex designs, such as the 'layer cake' and 'stalactite' designs, are each designed with specific purposes in mind, from attempts to rehabilitate dead areas of coral reef to creating a surface on which to grow shellfish commercially.

Fig. 3. The standard reef ball design (from the Reef Ball Foundation, www.reefball.org).

Through a combination of specifically-designed materials, and the placing of such materials in environments in which commercial populations of certain species such as lobster exist, a situation beneficial to both the local environment and local fishing communities may be reached. As the reef balls come in a range of sizes, including that similar to the boulders

installed where required around offshore wind turbines, they should be relatively easy to adapt to ensure they also fit the purpose of reducing scour around the base of the turbines, thereby also satisfying the key engineering purpose for which scour protection is deployed.
However, as with any development, economics is a major factor in the design, planning and construction of offshore wind farms. With projects already costing millions of pounds to get into the water, additional costs for items such as the Reef Balls, when standard gravel or boulders are equally effective for the primary need, may not be easily approved by developers.
However, there may be a mid-point to the discussions, if a material was identified which was relatively cheap to purchase and install (compared to the specially-designed Reef Balls), as well as being able to function equally well as scour protection and increased habitat around the base of the turbine towers.
Materials commonly used in sea-wall construction, such as dolos blocks, tetrapods or concrete jacks, are built for strength, able to withstand large amounts of pressure, and also have unique shapes which lock in to each other, gradually shifting in the weeks after installation to form tight bonds with adjacent blocks. Using these materials would allow the creation of a wide number of niches, and increased surface area compared to boulders of a comparable size, thereby allowing greater potential for colonisation.
Another key element in the potential habitat creation by offshore wind farms and their associated infrastructure / scour protection is the argument that these areas may become unofficial marine protected areas (MPAs). Although fishing activity is not directly banned within the boundaries of many offshore wind farms, and in many is taking place successfully, some fishing gear is not conducive to the environment within the site boundary, such as dredging, which could lead to entanglement in the inter-array cables associated with the turbines. It is therefore possible that some offshore wind farm sites may have low levels of fishing taking place within them. Fayram and de Risi (2007) suggest that by creating an MPA in the area surrounding offshore wind farms, with limited entry to fishing activity (both commercial and recreational), it may be possible to provide circumstances which would be beneficial to a number of parties. It is noted that in some cases, oil platforms have acted as *de facto* MPAs due to prevailing currents and the platform themselves preventing the use of several types of fishing gear. If the same is true for offshore wind farms, then the wind farm owners would benefit due to reduced risk of damage from passing vessels, fishing groups could benefit from locally enhanced stocks, and the benthic and fish communities could benefit from reduced disturbance from fishing activity. Therefore, although the main aims of offshore wind power generation and MPA designation vary considerably, in some situations they may be complimentary (Fayram and de Risi, 2007).
Through the installation of offshore wind turbines, one of the key changes for the surrounding marine environment is the introduction of a new dimension in habitat terms. Many of the areas into which offshore energy generation is expanding is, for ease of construction, relatively flat seabed, with very few vertical elements such as reefs or cliffs. Therefore, the addition of the turbines and their foundations can add vertical habitat where before there only existed horizontal habitat for species to colonise.
Although it is impossible to physically increase the volume of water column already existing as habitat, and it could be argued that the installation of turbine towers actually removes a negligible amount of water in the area, the installation alters the form of the water column habitat available.

Despite turbines being up to almost 1km apart in some larger developments, the addition of the vertical habitat can act as shelter for some fish species, creating structure in an otherwise featureless open ocean. Therefore, an increase in the ecological 'usefulness' of the area is brought about, and as a result, its carrying capacity.

This distance between individual turbines will also determine whether, from a community perspective, the turbines are independent of each other, or are able to act as one large area of introduced habitat. This varies between species, with 1km being well within the range of larger, motile species such as cod, lobster and some crab species (Linley *et al*, 2007), but for smaller fish species, or benthic organisms, which develop where they settle, there is less likely to be mixing between turbines. Although the design of the turbine layout is heavily based on economics, to ensure maximum wind, and therefore energy yield, consideration of the biological perspective in the array design at an early stage, could increase the potential for habitat creation to be as effective as possible.

5. Future study requirements

The status of offshore wind energy generation as a relatively young industry has both positive and negative aspects for developing its potential in habitat creation.

With few long-term studies of the changes in abundance and diversity of species within wind farms available, due to the relatively low number of developments currently operational, there are few datasets to fully analyse for the potential habitat gain which have been discussed in this chapter.

This problem of lack of long-term data also exists within the field of 'standard' artificial reefs, with few study programmes running longer than a couple of years in order to establish the initial stages of colonisation and succession (Perkol-Finkel and Benayahu, 2005).

In their research, Perkol-Finkel and Benayahu (2005) returned to previously-studied artificial reefs in the Red Sea, to determine what further developments occur ten years after deployment. It was noted that despite their close proximity, and equivalent depths, the community structure and species diversity differed between the artificial reefs and neighbouring natural reefs, which had acted as control sites for the early stages of the comparative study. Similar results are reported where shipwrecks and adjacent reef areas have been studied, with higher species diversity on the natural reefs. Naturally, the age of an artificial reef, whether intentional or not, will greatly affect its community structure, as certain species can only recruit after initial settling species have increased the complexity of the surface, making it suitable for secondary species. In one long-term study, it was estimated that the development of benthic communities in Pacific temperate waters might take up to fifteen years (Aseltine-Neilson *et al.*, 1999).

These findings highlight the need to ensure surveys of already operational offshore wind turbines, and their associated scour protection and infrastructure, continue throughout the lifetime of the project, which can be up to fifty years (Centrica, 2009). The Perkol-Finkel and Benayahu (2005) study also highlights the fact that surveying the development of life on turbines alongside neighbouring natural communities will allow evaluation of the biological and environmental benefit of the turbines as artificial reef structures.

Despite the issue of the lack of long-term datasets, the 'youth' of the industry could also mean that any methods identified for increasing the benefits of offshore turbines in terms of habitat creation may still be incorporated into the design of future projects as they come into the detailed design phases, prior to construction, where appropriate. It is therefore even more important for the results of survey work which has been undertaken to be widely distributed and discussed, allowing any possible design adjustments to be made before the major Round 3 developments reach the turbine-selection stage.

Further study work should also be directed at the various types of material most commonly used for scour protection, where deployed. Calculations have already determined that the level of habitat created varies depending on the type of scour protection deployed around the base of offshore wind turbines (Wilson and Elliott, 2009), and this could be developed further to incorporate other variables. These calculations were based upon a single diameter for gravel and boulder scour protection; however with slight changes to the size of the material used, significant changes may be made to the area available for early colonizing species, which will in turn attract a wider range of species. Taking this further, combining the methods of scour protection used within a single development, could have an additional beneficial effect. By introducing gravelly substrate, rocky reef environment and sea grass environment into a predominantly sandy seabed area, habitat diversity will be significantly increased, with each habitat created bringing with it the various communities which inhabit them.

Detailed survey work of the colonisation and succession of species on a range of scour protection materials, in the field, will assist in demonstrating the potential that offshore wind farms have in creating viable habitat, as well as allowing countries to reach their renewable energy targets.

The usefulness of specially-designed reef materials, such as the Reef Ball, as scour protection, should be investigated. If these materials are able to perform the main role of scour protection, then their deployment around turbines may be particularly beneficial to the receiving marine environment.

Despite the many potential benefits which may occur as a result of habitat creation around offshore wind farms, there must also be some level of caution. The introduction of new habitats in environments where such habitats did not previously exist may also introduce new species into the area, outside of their usual ranges. In addition, there is the possibility for high concentrations of certain predatory species, such as starfish, to colonise the turbines in such high numbers that they may have a negative impact on existing communities. Therefore, future colonisation studies around offshore turbines and their associated infrastructure should take particular note of these new species, and any interactions which may be taking place with existing communities.

6. Summary

The expansion of offshore wind farm development has the potential to bring about great benefits. Not only will the increase in renewable energy generation help in the fight against climate change, but through the introduction of new habitats into the marine environment, turbines can also act as artificial reefs, potentially increasing both species and habitat diversity.

For true artificial reef design and installation, a number of key factors need to be considered, including geographical location, size, orientation, complexity, durability, type of material, surrounding substratum, proximity to natural habitats, depth and water conditions (Perkol-

Finkel and Benayahu, 2005). Only purposely-planned artificial reefs can satisfy the full range of requirements for a truly successful reef, encouraging full colonisation and succession sequences, and becoming a useful tool for conservation or restoration of existing habitats / stocks / communities; however, with a bit more planning at the early stages of development, it should be possible for the development of offshore wind energy to satisfy a number of these requirements, and thereby become at least partially successful at creating habitat around its tower and foundation.

To illustrate the importance of structures placed within the marine environment, when four small oil platforms were removed from Californian waters in 1996, over 2000 tons of marine life were removed from the platform legs, and disposed of in landfill sites onshore (California Artificial Reef Enhancement Programme's website). Therefore, it is important to consider the decommissioning of any offshore turbines even before they are installed. Although it may not be feasible from a navigational safety point of view to leave all foundation structures in place once the towers and nacelles have been removed, it may be possible to leave some foundations in place, for example as part of an MPA once the wind farm itself has been decommissioned and removed.

A key aspect of the habitat creation argument is to get the issue wider appreciation at a higher industry level. If the gains to both the ecology and economy of the surrounding marine environment are known and understood more widely by developers, regulators and other stakeholder groups, then they may be able to form part of early discussions and negotiations with regards to specific project design and construction methods. Survey and research results should be published with an eye as to how they can be further utilised and adapted, with greater emphasis on the broader range of conservation, commercial or recreational gains which could be achieved.

Economics is another major aspect in offshore wind farm generation, another reason why better understanding of all implications, positive and negative, is essential. As described previously, the potential additional cost required to take full advantage of the habitat-creation potential of offshore wind farms may prove too great to convince developers, mindful of costs and profits, to alter plans and designs for their projects, without absolute evidence as to the benefits. However, given the potential for enhancement of commercial stocks, or conservation of particular communities or species, perhaps there is the possibility for local councils, fisheries associations or nature conservation groups to become involved, 'sponsoring' the installation of targeted scour protection, given the benefits that could be expected.

In conclusion, there is a large body of evidence for the benefits of artificial reefs in the marine environment, both intentionally designed and placed, and otherwise. Studies have shown that the introduction of almost any structure into the oceans will result in the colonisation of that structure, and that in many cases, this brings about increased productivity, rather than simply aggregating life from adjacent areas.

This increased productivity has the potential to bring about further benefits from both conservation and commercial perspectives, depending on the area in which the turbines are being installed, and whether any commercial / sensitive species already exist locally. The use of targeted scour protection could increase the capacity to help particular species, for example, through the installation of boulder protection in an area with a strong local lobster fishery. Using specially-designed materials may increase this beneficial capacity even further.

However, as with all young industries, there is still a need for greater understanding of both the impacts and potential benefits of offshore wind farms, and how the habitat-creation potential around the turbines and other infrastructure can be fully taken advantage of. Therefore, the results of all post-construction surveys, such as those discussed briefly previously in this chapter, should be collated and reviewed in detail, to gain an understanding of how colonisation works on specific foundation types, in specific areas, taking into account the communities already in existence in the receiving environment. Incorporating further survey results as they become available will increase this understanding, and give a range of time-frames for the study.

With careful consideration and planning then, the installation of wind turbines into the marine environment has the capacity to help combat climate change, and bring about benefits for not only the communities which already exist in the area, but potentially, introduce new such communities, with their subsequent commercial and conservational benefits.

7. References

Ackermann, T; Soder, L. (2002) An overview of wind energy status. *Renewable and Sustainable Energy Reviews*, 6, 67-127.

Anthoni, F. (2006) Invasion of the parchment worm. Accessed at: www.seafriends.org.nz/indepth/invasion.htm. Last accessed 13 September 2010.

Aseltine-Neilson, D.A.; Bernstein, B.B.; Palmer-Zwahlan, M.L.; Riege, L.E.; Smith, R.W. (1999) Comparisons of turf communities from Pendleton artificial reef, Torrey Pines artificial reef, and a natural reef using multivariate techniques. *Bulletin of Marine Science*, 65(1), 37-57.

Beaurea of Ocean Energy Management, Regulation and Enforcement (2010) Artificial reefs: Oases for marine life in the Gulf. Accessed at: http://www.gomr.boemre.gov/homepg/regulate/environ/rigs-to-reefs/artificial-reefs.html. Last accessed 7 September 2010.

BWEA (2005) British Wind Energy Association Briefing sheet – Offshore wind. Accessed at www.bwea.com/pdf/briefings/offshore05_small.pdf. Last accessed 13 September 2010.

California Artificial Reef Enhancement Programme (CARE). Website accessed at: http://calreefs.org/. Last accessed 7 September 2010.

Centrica (2009) Race Bank Offshore Wind Farm: Environmental Statement.

Elliott, M. (2002) The role of the DPSIR approach and conceptual models in marine environmental management: An example for offshore wind power. *Marine Pollution Bulletin* 44, iii–vii.

EMU (2008a) Barrow Offshore Wind Farm: Monopile Ecological Survey. Report to Barrow Offshore Wind Ltd, December 2008.

EMU (2008b) Kentish Flats Offshore Wind Farm: Turbine Foundation Faunal Colonisation Diving Study. Report to Kentish Flats Ltd, November 2008.

Fayram, A.H.; de Risi, A. (2007) The potential compatability of offshore wind power and fisheries: An example using bluefin tuna in the Adriatic Sea. *Ocean and Coastal Management*, 50, 597-605.

Forward, G. (2005) The potential effects of offshore wind power facilities on fish and fish habitat. Algonquin Fisheries Assessment Unit, Ontario Ministry of Fisheries Resources. Found at ozone.scholarsportal.info Last accessed 18 August 2007.

Hiscock, K. (2009) Revealing the reef: marine life settling on the ex-HMS Scylla. Online presentation available at: www.marlin.ac.uk/learningzone/scylla. Last accessed 20 August 2010.

Jenson, A.; Collins, K.J.; Free, E.K.; Bannister, C.A. (1994) Lobster (*Homarus gammarus*) movement on an artificial reef: the potential use of artificial reefs for stock enhancement. *Crustaceana* 67, 198-212.

Jha, A. (2008) UK overtakes Denmark as world's biggest offshore wind energy generater. The Guardian Online. Accessed at: http://www.guardian.co.uk/environment/2008/oct/21/windpower-renewableenergy1. Last accessed 10 August 2010.

Kopp, D; Bouchon-Navaro, Y.; Louis, M.; Bouchon, C. (2007) Diel differences in the sea grass fish assemblages of a Caribbean island in relation to adjacent habitat types. *Aquatic Botany*, 87, 31-37.

Langhamer, O.; Wilhelmsson, D. (2009) Colonisation of fish and crabs of wave energy foundations and the effects of manufactured holes - a field experiment. *Marine Environmental Research*, 68, 151-157.

Linley, E. A. S.; Wilding, T. A.; Black, K. D.; Hawkins, A. J. S.; Mangi, S. (2007) Review of the reef effects of offshore wind farm structures and potential for enhancement and mitigation. Report from PML Applications Ltd. to the Department of Trade and Industry. Contract no. RFCA/005/0029P

Linnane, A., Mazzoni, D. and Mercer, J. P. (2000) A long term mesocosm study on the settlement and survival of juvenile European lobster *Homarus gammarus* in four natural substrata. Journal of Experimental Marine Biology and Ecology 249 pgs 51-64.

Løkkeborg, S.; Humborstad, O-B.; Jørgensen, T.; Vold Soldal, A. (2002) Spatio-temporal variations in gillnet catch rates in the vicinity of North Sea oil platforms. ICES Journal of Marine Science, 59, S294-S299.

Louisiana Department of Wildlife and Fisheries (2005) Accessed online at: http://www.wlf.louisiana.gov/fishing/programs/habitat/artificialreef.cfm. Last accessed 12 August 2010.

Perkol-Finkel, S.; Benayahu, Y. (2005) Recruitment of benthic organisms onto a planned artificial reef: shifts in community structure one decade post-deployment. *Marine Environmental Research*, 59, 79-99.

Perkol-Finkel, S.; Benayahu, Y. (2007) Differential recruitment of benthic communities on neighbouring artificial and natural reefs. *Journal of Experimental Marine Biology and Ecology*, 340, 25-39.

Pihl, L.; Baden, S.; Kautsky, N.; Ronnback, P.; Soderqvist, T.; Troell, M.; Wennhage, H. (2006) Shift in fish assemblage structure due to the loss of sea grass *Zostera marina* habitat in Sweden. *Estuarine, Coastal and Shelf Science*, 67, 123-132.

Wilhelmsson, D.; Malm, T.; Ohman, M.C. (2006) The influence of offshore wind power on demersal fish. *ICES Journal of Marine Science*, 65(5), 775-784

Wilson, C.A.; Van Sickle, V.R.; Pope, D.L. (1987) Louisiana Artificial Reef Plan; Louisiana Department of Wildlife and Fisheries, Technical Bulletin No. 41, November 1987.

Wilson, J.C.; Elliott, M. (2009) The habitat-creation potential of offshore wind farms. *Wind Energy*, 12(2), 203-212.

10

Perceived Concerns and Advocated Organisational Structures of Ownership Supporting 'Offshore Wind Farm – Mariculture Integration'

Gesche Krause, Robert Maurice Griffin and Bela Hieronymus Buck
[1]Leibniz Center for Tropical Marine Ecology (ZMT), Bremen
[2]Department of Environmental and Natural Resource Economics, University of Rhode Island
[3]Alfred Wegener Institute for Polar and Marine Science (AWI), Bremerhaven
[4]Institute for Marine Resources (IMARE), Bremerhaven
[5]University of Applied Sciences Bremerhaven, Bremerhaven
[1,3,4,5]Germany
[2]USA

1. Introduction

In Germany a major political incentive exists currently to install large offshore wind farms (Tiedemann, 2003; BMU/Stiftung Offshore Windenergie, 2007). The promotion of wind power especially in offshore regions is mainly driven by the policy to reduce dependence on conventional fossil energy resources as well as the need to reduce the environmentally harmful CO_2 loads. Offshore wind farms are defined here as a group of wind turbines in the same confined area used for production of electric power in the open ocean. Moving off the coast to the offshore, wind turbines are less obtrusive than turbines on land, as their apparent size and noise is mitigated by distance. Since water has less surface roughness than land (especially in deeper waters), the average wind speed is usually considerably higher over the open water. At present 47 project applications for wind farms in the Economic Exclusive Zone (EEZ) of the German North Sea and in the Baltic Sea are in the planning process (BSH, 2008) with a total number of wind turbines per farm ranging between 80 and 500 (Buck et al., 2008). The strong expansion of offshore wind farms in the marine environment of the North Sea increases the stress on sea areas that have formerly been used for other purposes, such as for fishery or shipping activities, or that are still seemingly free of human activity (Krause et al., 2003; Wirtz et al., 2003).

Hence, the emerging offshore wind industry is quickly becoming a large stakeholder in the offshore arena (Gierloff-Emden, 2002; Dahlke, 2002; Tiedemann, 2003). This has lead to conflicts of interest among the different user groups and has encouraged research on the prospects of integrating maritime activities under a combined management scheme as newcomers such as wind farms make for additional claims exclude other uses, such as wild-harvest fisheries. In this context, integrating marine aquaculture with designated wind farm areas might provide chances to combine two industries in the frame of a multiple-use concept (Buck et al., 2009). The term marine aquaculture, or mariculture, refers to aquatic organisms cultivated in brackish or marine environments. Offshore aquaculture indicates a culture operation in a frequently hostile open ocean environment exposed to all kinds of sea states as well as being placed far off the coast. Nowadays the increasing limitation of favourable coastal sites for the development of modern aquaculture which is evident in various countries such as Germany, the Netherlands, Belgium, as well as others, has spurned this move offshore (Buck & Krause, 2011). This spatial limitation is mainly caused by the high degree of protected nearshore areas and by the fact that regulatory frameworks that assign specific areas for aquaculture operations are diverse and still emerging (Krause et al., 2003). Thus, little room for the expansion of modern coastal aquaculture systems in nearshore waters remain. In contrast, the number of competing users within offshore regions is relatively low, hence favouring the offshore environment for further commercial development, such as offshore wind farming and open ocean aquaculture. Spatial regulations offshore are scarce so far and clean water can be expected (Krause et al., 2003; Buck et al., 2009).

This chapter examines possible motivations for, and methods of, forming and managing an integrated facility where mariculture production resides within the physical boundaries of an offshore wind farm. It does so from an organisational science point of departure and takes into account the broad literature on organisational science and the particular context of the North Sea. The chapter closes with a short summary on the probable strategies of governance for future potential integration of offshore 'wind farm – mariculture activities'.

2. Methods

Existing insights relating to the research questions above are yet limited. Thus, an exploratory or discovery-oriented approach was chosen, in which the primary stipulation was that the research should be empirical. The results and deliberations presented here are generated from several focus group meetings, stakeholder workshops, and semi-structured interviews over the course of years of research on the subject of multi-use management of offshore wind farms and mariculture. The key findings are summarised in Buck (2002); Krause et al., (2003); Buck et al., (2008); Michler-Cieluch and Kodeih, (2007); Michler-Cieluch and Krause, (2008). Core of the discussions below are the findings from semi-structured interviews with people involved in the offshore wind farm sector and with individuals of the mussel fishery/farming sector in Germany.

Conclusions about suitable organisational structures are based on participants' views and their critical understanding of potential 'wind farm– mariculture integration'. The reason to focus primarily on these two actor groups is that they are potential adopters of such a multiple-ocean use scheme because of being the ones most directly involved in or affected by a possible organisational combination of the two working domains. Moreover, it is assumed that they are most knowledgeable about the particular offshore tasks and also aware of potential interferences between both sectors (Michler-Cieluch and Krause, 2008).

The findings are contextualized to the potential organisational structures and framework requirements expressed during interviews of personnel from the wind farm industry and mussel fishing/farming sector in which the issue of a multiple-use setting in the offshore realm was addressed. Altogether 34 semi-structured interviews were carried out, with most of the interviewees being engaged in operational or developmental activities of either sector.

However, different actors' relative power to bring about system change must be considered in investigating plausible future organisational structures. This also includes decisive legislative bodies that determine the specific constitutional rules to be used in crafting the set of collective-choice rules for multiple-use settings.

3. Results

The stakeholder analysis revealed that there are different types of actors involved in the offshore realm as in contrast to nearshore areas. Different types of conflicts, limitations and potential alliances surface. These root in the essential differences in the origin, context and dynamics of nearshore- versus offshore resource uses.

For instance, the nearshore areas in Germany have been subject to a long history of traditional uses through heterogeneous stakeholder groups of the local to national levels (e.g. local fisheries communities, tourism industry, port developers, military, etc.), in which traditional user patterns emerged over a long time frame. In contrast, the offshore areas have only recently experienced conflict. This can be attributed to the relatively recent technological advancements in shipping and platform technology, both of which have been driven by capital-strong stakeholders that operate internationally. Whereas there is a well-established organisational structure present among the stakeholders in the nearshore areas in terms of social capital and trust, as well as tested modes of conduct and social networks, these are lacking in the offshore area. Indeed for the latter, a high political representation by stakeholders is observed, that possess some degree of "client" mentality towards decision-makers in the offshore realm. These fundamental differences between the stakeholders in nearshore and offshore waters make a streamlined approach to multiple use management very difficult.

However, when addressing the identified offshore stakeholders, most of the interviewees were generally interested in this specific type of multiple-use setting and vitalized the conversation around the guiding questions with their own comments and ideas. Concurrently with judging 'wind farm – mariculture integration' as an idea worthy to consider, interviewees mentioned several framework requirements for initiating and effectively pursuing cross-sectoral offshore operation and organisation. Not only had certain preconditions to be fulfilled, for example the need to clarify the working tasks and siting of aquaculture installations in the forehand, but also overall regulatory conditions, e.g. determination of working rules, allocation of responsibilities, as well as commercial arrangements or actuarial regulations (Figure 1). The issue of sharing responsibilities in the context of everyday organisation and questions of ownership were especially stressed. In the following, we discuss the organisational structures of such multiple-use setting from an organisational perspective in more detail.

4. Discussion

The results of this stakeholder survey can help us to differentiate the likelihood of various mariculture-wind farm integration scenarios going forward, specifically regarding the various forms of ownership and management such a venture might take. The attitudes and perceptions of these groups prior to implementation are informed by their views on the possible synergies in production and organisational structure. Framing the results of the surveying and other contextual information in the well-developed literature of inter-firm organisation and cooperation will provide a basis for understanding the potential of this concept.

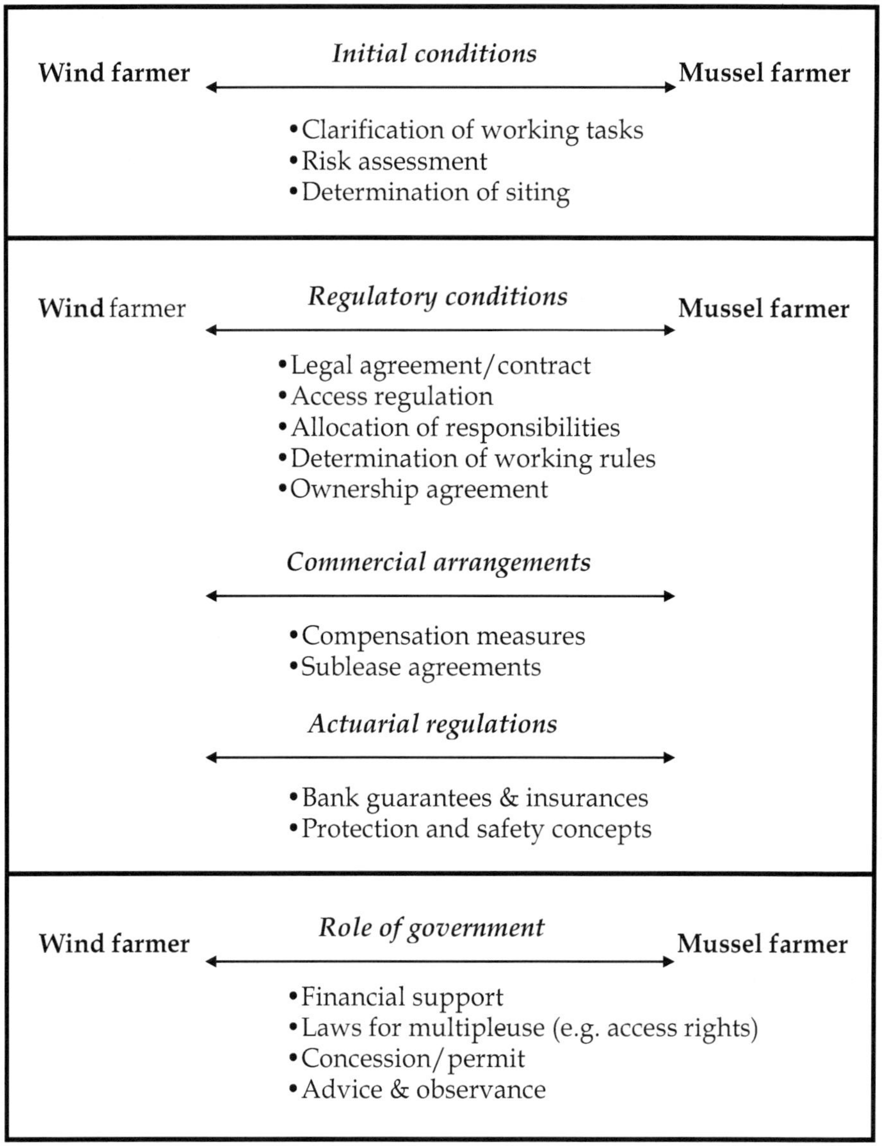

Fig. 1. Framework requirements for managing 'wind farm–mariculture integration' (modified after Michler-Cieluch and Krause, 2008).

The study of the formation of inter-firm organisations for the purpose of a mutually beneficial project or venture has roots in many research fields, with theories ranging from sociology, economics, psychology, business, and population ecology, amongst others (Osborn and Hagedoorn, 1997). The approach and methodology varies widely between these fields. Oliver Williamson has pioneered one economic approach, couching the study of governance and alliances in terms of transactions costs; see Williamson (1996) for a complete treatment. A related, but divergent approach is the work of Mark Granovetter, who takes the sociological concept of "social embeddedness" and uses it to justify the motivations and outcomes of inter-firm cooperation; see Granovetter (1985) for a review. The following analysis will incorporate, where possible, these related approaches and others to comprehensively view the challenges and potential of this new idea for offshore co-production.

4.1 Antecedent variables

There are many literature reviews that attempt to identify the basic elements necessary to conduct comparative research into inter-firm organisational structures and processes (Grandori & Soda, 1995; Osborn and Hagedoorn, 1997). Following Grandori and Soda's (1997) framework, the discussion will first identify the motives for cooperation between mariculturists and wind farmers and then look at some likely scenarios that may evolve for cooperation.

4.1.1 Production

A first motivation compelling these groups to consider a cooperative venture is the cost savings that may be available through production complementarities. Offshore construction and operation is more expensive than nearshore or onshore facilities for both industries, due primarily to large transportation costs and variables associated with the unpredictable and high-energy environment of the North Sea. Available working days per year may only be as much as 100 in the German North Sea (Michler-Cieluch et al., 2009a). It is of mutual interest of both groups to reduce their potential operating costs by collaborating in this difficult environment.

As outlined by responses in the survey, logistical cooperation is of joint interest. The ability to coordinate personnel movement to make joint use of transportation capital is a potential cost-saving avenue for either firm. In an offshore setting there could be significant potential for economies of scale in transport. Marginal increases in vessel capacity (boat or helicopter) could provide for reduced joint transportation costs, if an equitable agreement could be made for funding that capacity expansion.

It is worth noting that the operations and maintenance schedule of both offshore facilities will need to be highly coordinated internally, dictated by servicing schedules and operational tasks unique to each facility. Interlacing these schedules and any jointly used assets would however likely raise the costs of coordination, partly offsetting any gains made through complementary logistical planning.

There exists potential for other complementarities that may reduce costs for both firms in an integrated mariculture-wind farm facility and provide a motive for coordination:

- Interaction at the initial stages of planning and throughout the operating lifetime of the facility may possibly shorten the duration of the adaptive learning process that occurs in many businesses employing new technology or methods (Inkpen 2008; Nielsen 2010). The experience each group brings to the venture may provide a two-way information transfer that may improve each firm's technical efficiency of production. These economies of experience and shared experience effects may lower the average cost of production for each firm over time at a faster rate than if operating alone (Henderson, 1974).

- The current regulatory framework in countries on the North Sea makes few, if any, allowances for simultaneous economic use of the ocean area allotted for wind energy production. However, a strong momentum exists on the EU level to implement multiple concurrent uses of ocean space within the new Marine Strategy Framework Directive. In the event that these laws permit such activity in the future, there may exist an opportunity to reduce costs related to bureaucratic requirements and payments. For instance, if a given area was required to be leased from the government, the two firms may be able to split the cost of leasing. Similar logic applies to splitting the cost of pre-construction environmental studies and perhaps even engineering and other pre-construction plans. Cost savings may be offset by the extent to which these projects become more costly by including an expanded suite of activities.
- Current regulations in some North Sea countries also require insurance for offshore wind farms (Baugh, 2009). Dependent on the structure of the inter-firm agreement and the extent of the policy coverage, there may be an opportunity to hedge risk and lower insurance premiums versus operating independently at different sites. The extent to which this is possible is, in one way, determined by the economic viability of a joint operation and its associated organizational structure in the first place. As this is the focus of this paper and concurrent research on the economic feasibility of a joint mariculture-wind farm facility (Griffin and Krause, 2010), a more rigorous treatment of insurance is beyond the scope of this chapter.

The first, and most obvious, motive when looking at an inter-firm agreement from the vantage point of a mariculture firm is the ability to locate their operations in a protected offshore environment. Wind farms may be able to provide some safety for mariculture activities as well as provide a foundation for anchoring infrastructure (James and Slaski, 2006). One of the largest challenges to moving mariculture offshore is being able to protect it from the impacts of these high-energy environments (Bridger & Costa-Pierce, 2003). Recent development of innovative culturing devices for seaweed, mussels and fish (Buck and Buchholz, 2004; James & Slaski, 2006; Buck et al., 2006, Buck, 2007) within the offshore setting and particularly in wind farms may provide a cost benefit in installation and maintenance of infrastructure versus a stand-alone offshore farm.

Michler-Cieluch et al. (2009a) and Buck et al. (2008) suggest some other advantages that may reduce costs to mariculture firms:

- The offshore area provides a high quality environment for culturing the likely first candidates for offshore aquaculture, with high water quality, good oxygen conditions, less pollution, and less eutrophication than nearshore sites. This suggests that to meet a similar yield offshore may cost less due to superior growing conditions.
- The co-use of service platforms offshore may allow for more cost-effective maintenance and servicing. Dependent on the arrangement, personnel, equipment, or vessels may optionally have access to the service platform, providing flexibility in servicing and harvesting amongst other possibilities.
- James and Slaski (2006) mention that direct access to electrical power could allow for increased photoperiod production and higher levels of automation and remote operation.

- A first insight into the commercial benefit of a multiple-use scenario with aquaculture in offshore wind farms was calculated for a suspended mussel cultivation enterprise as a case study in the Germen North Sea (Buck et al., 2010).

The decision to partner with mariculture firms may also be motivated by cost considerations for wind energy firms. In an offshore setting where many users are competing for space, allowing the concurrent use of a wind farm for mariculture may provide a dual benefit to wind energy producers. First, depending on the form of cooperation, the wind energy firm may receive some level of direct compensation from the mariculture firm. This may come in the form of shouldering common costs, or be a direct stream of income as a "rental" rate, amongst other possibilities.

Secondly, the wind energy firm may experience a reduction of conflict with other users (James and Slaski, 2006). An integrated facility will likely not allow other users to enter that space which could jeopardize the safe operation of a heavily utilized offshore area (Mee, 2006). A corollary to this is that an integrated facility could be perceived as a sign of good faith and cooperation by wind energy producers in the often contentious sociopolitical landscape of exclusionary utilization of offshore commonly held resources. To date, the offshore wind farm operators hold "client" ties with the decision-makers, in which other users and their interests are not included in development considerations. By finding solutions which could be perceived as "win-win" for multiple stakeholders in the offshore setting, the wind energy operator may improve their public perception (Gee, 2010). In turn this may have the positive economic impact of reducing their political risk and potentially their cost of financing and insurance premiums.

This is not to suggest that there is only upside for a wind energy firm in collaborating. It is possible that they may experience a reduction in flexibility to engage in infrastructure projects as a result of inflexible growing seasons on the mariculture side (Mee, 2006). Taking on mariculture to the exclusion of shipping, wild harvest fisheries, or other interests may still result in alienation and political risk if excluded parties are not granted concessions elsewhere. It may also be the case that the transaction cost of implementing a joint agreement may be high enough to discourage entering into such an agreement. Flexibility in changing the collaborative arrangement as production strategies are adapted may encourage cost savings (Grandori and Soda, 1995), but may also be more costly to initially build into the agreement.

The motivations cited above are descriptive in nature and do not endeavour to model or quantify the interactions or the nominal values of these factors. As a set of potential cost savings from complementary production activities, they make a case for exploring additional motivations for collaboration.

4.1.2 Organisational coordination

4.1.2.1 Research

Grandori and Soda (1995) point out that collaboration is often motivated by reductions in governance costs and other factors unique to the industries or to the context in which the agreement is made. This section will first describe predictors from the literature which may support or impede collaboration, then will address related themes from each industry.

There has been extensive research into the pre-agreement predictors of collaboration, and the ongoing success of this collaboration. These can be related to the role of the respective asset portfolio. In this context of considerable natural resource dependency, the capital assets (natural, physical, human, financial and social capital), the activities, and the access to these (mediated by institutions and social relations) determine the income and the "livelihood platform" of users of natural resources (Niehof, 2004; Bond et al., 2007). Capital assets are not only resources that people use in building livelihoods, they are assets that give them the capability to be and to act (Badjeck, 2008).

In particular, asset-specificity is thought to be an important predictor of whether or not an inter-firm collaboration will emerge (Williamson, 1981; Grandori and Soda, 1995). Asset specificity is defined as the extent to which the investments made to support a particular transaction have a higher value than they would have if they were redeployed for any other purpose (McGuinness, 1994). In a successful agreement, bilaterally held assets and rights would be clearly specified, as well as the specific conditions under which the agreement could take place. This should prevent an opportunistic change of strategy by either party.

Of special importance in this case is site-specificity. The mariculture firm is specifically looking to gain the right to produce at an offshore wind energy site; this is the most essential piece to an agreement. A successful agreement must convey secure access to these rights foremost, and also clearly delineate any other joint assets or rights. In some areas of the North Sea where suitable alternative sites for mariculture are difficult to find, this may be even more important. Contracts should be comprehensive enough to avoid creating an incentive structure which undercuts the initial reasons for cooperation. The complexity of the joint agreement is also affected by additional interdependencies which refine the nature of the assets exchanged (Obsorn & Baughn, 1990, Bond et al., 2007).

The degree of differentiation between firms is a strong predictor of inter-firm coordination. This includes the distance among the objectives and orientations of these firms, as well as psychological differences in cognitive and emotional processes. It is interesting to note that while an excessive degree of differentiation in this regard has been identified as a cause of bureaucratic failure and disintegration of firms in the literature, diversity of resources controlled by the collaborating parties is considered a successful predictor of cooperation (Grandori and Soda, 1995). Williamson (1981) stated that "there are so many different types of organisations because transactions differ so greatly and efficiency is only realized if governance structures are tailored to the specific needs of each type of transaction."

Even more, Granovetter (1985) argues that all economic relations between firms occur in a broader social context, and this "embeddedness" plays a strong hand in market outcomes. Social and market conditions at the time of agreement may change the nature of the agreement or preclude the possibility altogether. The next section will discuss the context and common views held by the primarily affected stakeholders.

4.1.2.2 Context and Views

The mariculture industry in the North Sea has historically been concentrated entirely in the nearshore areas. Increasing competition from shipping, energy facilities, and conservation initiatives has added to pressure from wild harvest fisheries to constrain or reduce the available area for cultivation (CWSS, 2002; Michler-Cieluch et al., 2009b). Of the countries poised to make major commitments in the near term to offshore wind energy in the North Sea, there is not a particularly strong mariculture sector. That is the case in England, which has experienced significant offshore wind development already, though there is a well-developed salmon rearing industry in Scotland. Currently, no significant mariculture operations are being conducted outside of 12 nautical miles in Germany (Michler-Cieluch et al., 2009a), and there is considerable doubt about whether appropriate equipment and technology is available to do so

(Mee, 2006). There has been some consolidation across the industry in this area, especially among the salmon producers (James and Slasksi, 2006), but these businesses have a relatively small capitalization in comparison to wind energy developers.

Studies have shown that seaweed and mussels could be the best candidates in an extensive culturing environment based on biological, engineering, and economic considerations (Buck, 2002; Buck and Buchholz, 2005; Buck et al., 2008); this has the added advantage of being seen as "fitting in" in an environmentally and socially responsible manner over fish culture (RICRMC, 2010). Finfish cultured at offshore sites may have more economic potential in the market, but could have larger direct costs due to the intensive nature of culturing in a remote location (RICRMC, 2010) and are potentially more controversial from an environmental point-of-view.

On the other side, the offshore wind energy sector is rapidly developing in the North Sea. The UK, Denmark, and Germany have the most extensive development in terms of installed capacity or farms in varying stages of development (EWEA, 2009a). In the next ten years, the European Wind Energy Association expects the offshore capacity to quadruple in Europe (EWEA, 2009b). Currently the industry is still in an early stage, and projects still face a considerable amount of risk and uncertainty. The financial capital required to enter this business is large, and hence this industry is populated by developers who are backed by large utilities and consortiums of banks, utilities, and other conglomerates such as General Electric and Siemens.

The wind energy industry has the support of governments across the North Sea, and is seen as part of the solution in switching to a new "green" energy economy. Subsidies and favourable regulatory status have propelled the creation of offshore wind farms (Snyder and Kaiser, 2009), possibly to the detriment of other ocean users (Mee, 2006). There are also continued concerns about the environmental impact of wind farms on the adjacent ecosystem throughout its lifecycle, particularly on adjacent marine life and migrating birds (RICRMC, 2010).

It is against this backdrop in which agreements on an integrated wind energy-mariculture facility could be made. Prior beliefs held by firms going into the agreement process may play a large role in the success of those negotiations. In constructing the following generalizations about the viability of a collaborative agreement, the results of Mee (2006), Michler-Cieluch and Krause (2008), and Michler-Cieluch et al. (2009), are referenced. In general:

- Both groups have little interest in the joint-planning process, and have uncertain assessments of mutual gains from cooperation.
- In the case of deep-water offshore farms, the distance from shore does not foster cooperation. If these facilities were closer to shore it would make the economics more compelling for both groups.
- There are divergent interests in the resource system and perceptions of management problems (Michler-Cieluch and Kodeih, 2007).
- The lack of personnel with cross-sector experience makes it difficult for either group to envision how an integrated facility could work.
- No prior formal or informal relations between the two groups may hinder coordination (Grandori and Soda, 1995; Fukuyama, 1995).
- The relative net revenue disparity between operations is so large as to provide little incentive for a wind farm to engage in a collaborative project (Griffin and Krause, 2010).
- Doing business in the offshore area is environmentally and technically challenging. With a predilection towards risk, these groups may be in a unique position for collaboration where other investors and businesses would not be interested.

New industries face significant challenges in establishing themselves as legitimate. Stakeholders, policymakers, and others in the market will not be fully convinced of the viability of this concept until there is comprehensive organisational legitimacy (Yeow, 2006). As this is a new industry concept, it is not surprising that there could be significantly divergent interests and marked uncertainty regarding initial and subsequent viability. The experience in either industry is limited, and a collaborative effort has no precedent.

4.2 Modes of cooperation

Any analysis of likely management scenarios for an integrated wind energy-mariculture facility should include a discussion of the relevant government policy. In the North Sea area, this concept is ahead of the current regulatory system in place. So far, no systematic regulations exist addressing this multi-use concept in the context of industry support. While current legislation may preclude concurrent economic activity within offshore wind farms, that likely stands as a *de facto* law absent any regulatory consideration on the matter. Given the strong push for spatial efficiency and multi-use concepts in the maritime waters in the EU and elsewhere (Krause et al., 2003; Lutges and Holzfuss, 2006), it is likely that more comprehensive regulatory frameworks will develop shortly. There are three likely avenues under which an integrated mariculture-wind energy facility may be organized. These are not exhaustive, or mutually exclusive from each other, but rather provide a straightforward method for categorizing potential outcomes.

4.2.1 Sole owner

At one polar extreme, a multiple use business plan could be enacted by a sole company without any cooperation. In all likelihood, this fits better from the direction of the wind energy producer, who would have easier access to the financial resources needed. The aforementioned complexity of drafting and following a contract with an outside firm may make this an appealing choice. Governance structures that have better transaction cost economizing properties are preferable from an economic point of view, and transaction cost economics suggests that full vertical integration completely resolves issues related to hold-ups and misaligned incentives (Williamson, 1981; Williamson, 1979; Johnson and Houston, 2000).
Considering that the area occupied by wind turbines is roughly 1-3% of the total area of an offshore wind farm (Mee, 2006), the potential for further net revenue via mariculture may be alluring to a wind energy firm. Economies of scope, i.e. simultaneously producing two products with a lower average cost than if undertaken separately, may provide the financial catalyst. Current research is assessing the economic merits of a joint mariculture-wind energy facility and will help illuminate the viability of such a venture from multiple perspectives (Griffin and Krause, 2010).
As an economic decision, undertaking this as a sole firm partly rests on the ability of the wind energy producer to culture products at a similar or lower average cost than if they had negotiated a contract or formed a joint venture with a firm who specializes in mariculture. A major impediment to this scenario is the lack of technical capacity and experience to extend the scope of production into offshore mariculture. Thus, while a sole ownership approach may initially appear promising, the degree of risk involved in operating two very different businesses at the same location is high. The degree to which personnel with specialized knowledge could be brought in to oversee and conduct these operations would likely dictate the relative risk of internalizing both productive activities.

4.2.2 Negotiated contract

Robinson (2008) found that, on average, alliances occur more in riskier industries than do internal projects, and hence alliances are used to organize activities that are riskier than a firm's average inside project. Expanding to an industry-level analysis, he found that alliance intensity across industries is positively associated with the risk difference between the two industries. This dynamic could play an important role in alliance formation versus single firm management of a multi-use facility. Negotiated contracts are another way an integrated facility might be managed and risk distributed. Contracts could take a variety of forms, such as a joint venture or a consortium or any form of subcontract. The key tenets here are that the outlined interdependence between firms must provide benefit to each party (Pareto-improving) and be perceived as fair by the participating entities. Continued cooperation between parties must be sustainable by the underlying incentive structure (Grandori and Soda, 1995). The potential of coordination can be large when firms coordinate core skills to form an alliance with unique capabilities that neither partner could efficiently provide alone. Michler-Cieluch and Krause (2008) showed that there is sufficient scope for such wind farm-mariculture cooperation in terms of operation and maintenance activities.

The process of drawing up a contract that delineates the lines of cooperation between firms is fraught with challenges. Hold-up hazards increase when complexity and uncertainty make writing and enforcing contracts difficult (Williamson, 1979), and when products require asset-specific investments, two conditions that hold in this case. Economic efficiency compels firms to engage in integrated organisational structures over simple contracts or sole ownership only when there are offsetting benefits to doing so (Johnson and Houston, 2000). These could fall under any of the previously outlined benefits from cooperation, such as reduced production costs, organisational efficiencies, or pooling risk - but these benefits are not guaranteed. Nielsen (2010) argues that all alliance contracts are necessarily incomplete because of the parties' inability to write an *a priori* comprehensive agreement that covers all future contingencies, and thus these contracts may enhance or prohibit desired outcomes. Therefore, in order to be successful, all stakeholders involved in such joint cooperation agreements must be informed and clear about their expectations, rights and the duties involved.

There is considerable research regarding the predictors of success in joint ventures and other alliances. Johnson and Houston (2000) find that only joint ventures between firms in related businesses are likely to generate operating synergies, and that combinations of dissimilar firms can reduce value by contributing to bureaucracy and lack-of-focus. Beamish (1994) finds that the good intentions and rational motives behind alliances are often not congruent with the strategic direction of either firm on its own, and can lead to poor performance and instability. In the case where firms with asymmetric resource endowments enter into a joint venture, Kumar (2007) finds that asymmetric wealth gains arise via the negative wealth transfer effects of resource appropriation by the firm with more valuable resources. Lastly, Michler-Cieluch et al. (2009a) suggest that initial collaborative research between sectors prior to the design and execution of a commercial agreement is mandatory.

In the case of the wind farm-mariculture topic, our interviews and survey work suggests that the stakeholders in a potential mariculture-wind energy facility may be amenable to some type of contracted agreement. There exists some interest in a prior joint research initiative and feasibility study, and respondents have suggested that they would be open to the idea of contracting out culturing activities at the site of an offshore wind farm. It does seem unlikely though at this point that a contracted solution could occur in the absence of some intervening third body (Michler-Cieluch, 2009a). However, an advisory or some other external group helping to coordinate and mediate generally improves the chances of reaching a successful agreement (Noble, 2000).

4.2.3 Legislated

In the case where finding a market solution for multi-use in the offshore setting is not possible, a legislative prescription can still attain desired policy goals of spatial efficiency in the ocean area. The use of mandates, subsidies, tariffs, and other policy tools can change the incentives of the current economic environment to make the multi-use concept economically viable.

As there is a growing focus on coastal zone management and the efficient and equitable use of coastal resources in the EU, US, and elsewhere (Krause et al., 2003, Lutges and Holzfuss, 2006; RICRMC, 2010), policy makers may find policy instruments as a palatable solution for achieving policy goals. Mariculture can offer expanded employment opportunities to rural peripheral regions and displaced fishermen in the area of a wind energy facility, and potentially make wild harvest fisheries more productive if mariculture areas act as nurseries for wild fish (Mee, 2006). Indeed, multi-use layering of economic activities can maximize the value of offshore resources while reducing conflict between stakeholder groups. Promoting a multi-use concept would not be an uncommon step; regulators have already shown that they are comfortable with using legislation to spur growth in the offshore wind energy industry.

A clear, coherent, and stable regulatory framework is a bare minimum when firms make financial decisions in the inherently risky offshore marine environment. Managers need to be able to predict with some certainty the expected outcomes of changes in strategy, be it an internal decision or the decision to form an external alliance. Carroll et al. (1988) have found that fragmentation in the structure of State decision making is shown to lead to more elaborate and costly inter-organisational networks. The decision to actively foster cooperation on a multi-use concept should largely be dependent on market conditions, and the potential social benefits available from multi-use facilities.

5. Conclusion

The discussion thus far has attempted to frame the potential cooperation in a multi-use setting in the context of the broader social, political, and economic spheres, while also illuminating the perceptions and characteristics of the particular industries themselves. It appears clear that uncertainty and risk are large components of this discussion, and naturally were brought up by survey respondents. The likelihood and form of collaboration in the near future will be shaped by how well this risk and uncertainty is addressed.

It is apparent that the orchestration of a multi-use concept such as an integrating wind energy and mariculture will be difficult. First results indicate that practical multifunctional use of offshore areas requires technical and economic feasibility as a basic prerequisite to assure that both offshore wind farm operators and mariculturists will support a multi-use concept. This suggests that as more information emerges on the economic and technical viability of this, it will be clearer if this is a practical approach towards rationalizing marine stewardship in the offshore setting. Concurrent to this, it will fall to policy-makers to sanction the range of options for how such a facility might be managed. The discussion here is meant to enlighten the debate going forward on the relative merits of various management alternatives, while also illuminating the motivations for cooperation from a business standpoint.

6. References

Badjeck, M.C. (2008). Vulnerability of coastal fishing communities to climate variability and change: implications for fisheries livelihoods and management in Peru. PhD Thesis, University of Bremen, 227.

Baugh, M. (2009). Insurance: Supporting the Wind Energy Sector, In: *The Offshore Wind Revolution: Making Sense of the Opportunity*, Morrison, R. (Ed.), 63-72, Reuters Project Finance International.

Beamish, P. (1994). Joint Vetures in LDCs: Partner Selection and Performance. *Management International Review*, 34, 2, 60-74.

BMU/Stiftung Offshore Windenergie (2007) Offshore wind power deployment in Germany. Federal Ministry for the Environment, Nature Conservation and Nuclear Safety (eds).

Bond, R., Kapondamgaga, P.H., Mwenebanda, B., Yadav, R.P.S., Rizvi, A. (2007). Monitoring the livelihood platform: reflections on the operation of the Livelihood Asset-Status Tracking method from India and Malawi. Impact Assessment and Project Appraisal, 25, 4, 301-315.

Bridger, C., & Costa-Pierce, B. (2003). *Open Ocean Aquaculture: From Research to Commercial Reality*. Proceedings of the 2003 World Aquaculture Society Conference, Baton Rouge, LA.

BSH. (2008). Wind farms. Bundesamt für Seeschifffahrt und Hydrographie (Federal Maritime and Hydrographic Agency). Hamburg/Rostock, Germany. (Available March 2008 at http://www.bsh.de)

Buck, B. H. (2002). Open Ocean Aquaculture und Offshore Windparks. Eine Machbarkeitsstudie über die multifunktionale Nutzung von Offshore-Windparks und Offshore-Marikultur im Raum Nordsee. *Reports on Polar and Marine Research*, Alfred Wegner Institute for Polar and Marine Research, Bremerhaven, Germany.

Buck, B.H. (2007). Experimental trials on the feasibility of offshore seed production of the mussels *Mytilus edulis* in the German Bight: Installation, technical requirements and environmental conditions, *Helgoland Marine Research* 61, 87-101.

Buck, B. H. & Buchholz, C. M. (2004). The Offshore-Ring: A New System Design for the Open Ocean Aquaculture of Macroalgae. *Journal of Applied Phycology*, 16, 355-268.

Buck, B. H. & Buchholz, C. M. (2005). Response of Offshore Cultivated *Laminaria saccharina* to Hydrodynamic Forcing in the North Sea. *Aquaculture*, 250, 95-122.

Buck, B. H. & Krause, G. (2011) Integration of Aquaculture and Renewable Energy Systems. Encyclopedia of Sustainability Science and Technology. In press

Buck, B. H., Berg-Pollack, A., Assheuer, J., Zielinski, O. & D. Kassen. (2006). Technical Realization of Extensive Aquaculture Constructions in Offshore Wind Farms: Consideration of the Mechanical Loads, Proceedings of the 25th International Conference on Offshore Mechanics and Arctic Engineering, OMAE 2006 : presented at the 25th International Conference on Offshore Mechanics and Arctic Engineering, 4-9 June 2006, Hamburg, Germany / sponsored by Ocean, Offshore, and Arctic Engineering, ASME. New York, NY : American Society of Mechanical Engineers, pp 1-7.

Buck, B. H., Krause, G., Michler-Cieluch, T., Brenner, M., Buchholz, C., Busch, J., Fisch, R., Geisen, M., & Zielinski, O. (2008). Meeting the Quest for Spatial Efficiency: Progress and Prospects of Extensive Aquaculture within Offshore Wind Farms. *Helgoland Marine Research*, 62, 3, 269-281.

Buck, B. H., Ebeling, M. W., & Michler-Cieluch, T. (2010). Mussel cultivation as a co-use in offshore wind farms: potential and economic feasibility. Aquaculture Economics and Management 14(4): 1365-7305.

Carroll, G., Goodstein, J., & Gyenes, A. (1988). Organisations and the State: Effects of the Institutional Environment on Agricultural Cooperatives in Hungary. *Administrative Science Quarterly*, 33, 2, 233-256.

Dahlke, V. (2002). Genehmigungsverfahren von Offshore-Windenergieanlagen nach der Seeanlagenverordnung. Natur und Recht, 24, 472-479.

EWEA. (2009a). *Operational Wind Farms in Europe: End 2009*. Report by the European Wind Energy Association.

EWEA. (2009b). *Oceans of Opportunity: Harnessing Europe's Largest Domestic Energy Resource*. Report by the European Wind Energy Association.

Fukuyama, F. (1995). *Trust: The Social Virtues and the Creation of Prosperity*. Hamish Hamilton, London, UK.

Gee, K. (2010). Offshore Wind Power Development as Affected by Seascape Values on the German North Sea Coast. *Land Use Policy*, 27, 2, 185-194.

Gierloff-Emden, H.G.R. (2002). Wandel der Umwelt der See- und Küstenlandschaft der Nordsee durch Nutzung von Windenergie. Mitteilungen der Österreichischen Geographischen Gesellschaft, 144, 219-226.

Grandori, A. & Soda, G. (1995). Inter-firm Networks: Antecedents, Mechanisms and Forms. *Organisational Studies*, 16, 2, 183-214.

Granovetter, M. (1985). Economic Action and Social Structure: The Problem of Embeddedness. *American Journal of Sociology*, 91, 3, 481-510.

Griffin, R. & Krause, G. (2010). Economics of Wind Farm-Mariculture Integration. Working Paper, Department of Environmental and Natural Resource Economics, University of Rhode Island.

Henderson, B. (1974). The Experience Curve Reviewed vs. Price Stability. *Perspectives*, No. 149, Boston Consulting Group.

Inkpen, A. (2008). Knowledge Transfer and International Joint Ventures: The Case of NUMMI and General Motors. *Strategic Management Journal*, 29, 447-453.

James, M.A. & Slaski, R. (2006). Appraisal of the Opportunity for Offshore Aquaculture in UK Waters. Report of Project FC0934, commissioned by Defra and Seafish from FRM Ltd., 119 pp.

Johnson, S. & Houston, M. (2000). A Reexamination of the Motives and Gains in Joint Ventures. *Journal of Financial and Quantitative Analysis*, 35, 1, 67-85.

Krause, G., Buck, B.H., Rosenthal, H. (2003). Multifunctional Use and Environmental Regulations: Potentials in the Offshore Aquaculture Development in Germany. Proceedings of the Multidisciplinary Scientific Conference on Sustainable Coastal Zone Management "Rights and Duties in the Coastal Zone", 12-14 June 2003. Stockholm, Sweden.

Kumar, M.V. (2007). Asymmetric Wealth Gains in Joint Ventures: Theory and Evidence. *Finance Research Letters*, 4, 19-27.

Lutges, S. & Holzfuss, H. (2006). Integrated Coastal Zone Management in Germany: Assessment and Steps Towards a National ICZM strategy. German Federal Ministry for the Environment, Nature Conservation, and Nuclear Safety, 12 pp.

McGuinness, T. (1994). Markets and Managerial Hierarchies. In: *Markets, Hierarchies, and Networks*, Thompson, G. et al. (Eds.), 66-81, Sage, London, England.

Mee, L. (2006). Complementary Benefits of Alternative Energy: Suitability of Offshore Wind Farms as Aquaculture Sites. Report of Project 10517, commissioned by Seafish, 36 pp.

Michler, T. & Kodeih S. (2007). Offshore wind energy. Integration of Mariculture in Offshore Wind Farms. *Coastline Magazine*, 16, 1, 8 pp.

Michler-Cieluch, T., Krause, G. (2008). Perceived Concerns and Possible Management Strategies for Governing 'Wind Farm-mariculture Integration'. *Marine Policy*, 32, 6, 1013-1022.

Michler-Cieluch, T., Krause, G., & Buck, B. (2009a). Reflections on Integrating Operation and Maintenance Activities of Offshore Wind Farms and Mariculture. *Ocean & Coastal Management*, 52, 57-68.

Michler-Cieluch, T., Krause, G., & Buck, B. (2009b). Marine Aquaculture within Offshore Wind Farms: Social Aspects of Multiple Use Planning. *GAIA*, 18, 2, 158-162.

Nielsen, B. (2010). Strategic Fit, Contractual, and Procedural Governance in Alliances. *Journal of Business Research*, 63, 682-689.

Noble, B. (2000). Institutional Criteria for Co-management. *Marine Policy*, 24, 1, 69-77.

Osborn, R. & Baughn, C. (1990). Forms of Interorganisational Governance for Multinational Alliances. *The Academy of Management Journal*, 33, 3, 503-519.

Osborn, R. & Hagedoorn, J. (1997). The Institutionalization and Evolutionary Dynamics of Interorganisational Alliances and Networks. *The Academy of Management Journal*, 40, 2, 261-278.

Pauly, D., Christensen, V., Guénette, S., Pitcher, T.J., Sumaila, R.U., Walters, C.J., Watson, R., & Zeller, D. (2002). Towards Sustainability in World Fisheries. *Nature*, 418, 689–695.

Rhode Island Coastal Resources Management Council. (2010). *Rhode Island Ocean Special Area Management Plan*. 1856 pp.

Robinson, D. (2008). Strategic Alliances and the Boundaries of the Firm. *The Review of Financial Studies*, 21, 2, 649-681.

Snyder, B. and Kaiser, M. (2009). A Comparison of Offshore Wind Power Development in Europe and the U.S.: Patterns and Drivers of Development. *Applied Energy*, 86, 1815 56.

Tiedemann, A. (2003). Windenergieparke im Meer - Perspektiven für den umweltverträglichen Einstieg in eine neue Großtechnologie. In: Lozán J, et al. (eds). Warnsignale aus Nordsee & Wattenmeer: Eine aktuelle Umweltbilanz. Wissenschaftliche Auswertungen, Hamburg, 142–148.

Williamson, O. (1979). Transaction-Cost Economics: The Governance of Contractual Relations. *Journal of Law and Economics*, 22, 2, 233-261.

Williamson, O. (1981). The Economics of Organisation: The Transaction Cost Approach. *American Journal of Sociology*, 87, 548-577.

Williamson, O. (1996). *The Mechanisms of Governance*. Oxford University Press, ISBN 0-19-507824-1, New York, New York.

Wirtz, K.W., Tol, R.S.J., & Hooss, K.G. (2003). Mythos "Offene See": Nutzungskonflikte im Meeresraum. In: Lozan, L. et al (eds). Warnsignale aus Nordsee & Wattenmeer. Eine aktuelle Umweltbilanz. Wissenschaftliche Auswertungen, Hamburg, 157-160.